Advanced Structured Materials

Volume 41

Series Editors

Andreas Öchsner
Lucas F. M. da Silva
Holm Altenbach

For further volumes:
http://www.springer.com/series/8611

Andreas Öchsner · Holm Altenbach
Editors

Experimental and Numerical Investigation of Advanced Materials and Structures

Springer

Editors
Andreas Öchsner
Faculty of Biosciences and Medical
 Engineering (FBME)
University of Technology Malaysia - UTM
Skudai, Johor Bahru
Malaysia

and

Faculty of Engineering and Built
 Environment
The University of Newcastle
Australia

Holm Altenbach
Lehrstuhl für Technische Mechanik
Institut für Mechanik, Fakultät für
 Maschinenbau
Otto-von-Guericke-Universität Magdeburg
Magdeburg
Germany

ISSN 1869-8433 ISSN 1869-8441 (electronic)
ISBN 978-3-319-00505-8 ISBN 978-3-319-00506-5 (eBook)
DOI 10.1007/978-3-319-00506-5
Springer Cham Heidelberg New York Dordrecht London

Library of Congress Control Number: 2013945162

© Springer International Publishing Switzerland 2013

This work is subject to copyright. All rights are reserved by the Publisher, whether the whole or part of the material is concerned, specifically the rights of translation, reprinting, reuse of illustrations, recitation, broadcasting, reproduction on microfilms or in any other physical way, and transmission or information storage and retrieval, electronic adaptation, computer software, or by similar or dissimilar methodology now known or hereafter developed. Exempted from this legal reservation are brief excerpts in connection with reviews or scholarly analysis or material supplied specifically for the purpose of being entered and executed on a computer system, for exclusive use by the purchaser of the work. Duplication of this publication or parts thereof is permitted only under the provisions of the Copyright Law of the Publisher's location, in its current version, and permission for use must always be obtained from Springer. Permissions for use may be obtained through RightsLink at the Copyright Clearance Center. Violations are liable to prosecution under the respective Copyright Law.
The use of general descriptive names, registered names, trademarks, service marks, etc. in this publication does not imply, even in the absence of a specific statement, that such names are exempt from the relevant protective laws and regulations and therefore free for general use.
While the advice and information in this book are believed to be true and accurate at the date of publication, neither the authors nor the editors nor the publisher can accept any legal responsibility for any errors or omissions that may be made. The publisher makes no warranty, express or implied, with respect to the material contained herein.

Printed on acid-free paper

Springer is part of Springer Science+Business Media (www.springer.com)

Preface

Experimental and numerical investigation of materials and structures is nowadays an important discipline which enables a better and more reliable application of engineering components. Furthermore, limits of materials and structure can be accurately determined which may influence the design process and result, for example, in much lighter structures than a few decades ago. A lot of these advancements are connected with the increased computer power (hardware) and the development of well-engineered computer software. This directly influences the capability to bring novel advanced materials and structures to application. Only if the performance of new materials and structures can be sufficiently predicted and guaranteed, they will find their way in industrial applications.

The 6th International Conference on Advanced Computational Engineering and Experimenting, ACE-X 2012, was held in Istanbul, Turkey, from 1–4 July, 2012 with a strong focus on computational-based and supported engineering. This conference served as an excellent platform for the engineering community to meet with each other and to exchange the latest ideas. This volume contains 19 revised and extended research articles written by experienced researchers participating in the conference. The book will offer the state-of-the-art of tremendous advances in mechanical, materials, and civil engineering, ranging from composite materials, application of nanostructures up to automotive industry and examples taken from oil industry. Well-known experts present their research on damage and fracture of material and structures, materials modeling and evaluation up to image processing, and visualization for advanced analyses and evaluation.

The organizers and editors wish to thank all the authors for their participation and cooperation which made this volume possible. Finally, we would like to thank the team of Springer-Verlag, especially Dr. Christoph Baumann, for the excellent cooperation during the preparation of this volume.

April 2013

Andreas Öchsner
Holm Altenbach

Contents

Neural Model for Prediction of Tires Eigenfrequencies............ 1
Zora Jančíková, Pavel Koštial, Dana Bakošová, David Seidl,
Jiři David, Jan Valíček and Marta Harničárová

**Effect of Steady Ampoule Rotation on Radial Dopant Segregation
in Vertical Bridgman Growth** 13
Nouri Sabrina, Benzeghiba Mohamed and Ghezal Abdrrahmane

**Discontinuity Detection in the Vibration Signal
of Turning Machines**...................................... 27
Joško Šoda, Slobodan Marko Beroš, Ivica Kuzmanić and Igor Vujović

**Visualization of Global Illumination Variations
in Motion Segmentation**................................... 55
Igor Vujović, Ivica Kuzmanić, Joško Šoda and Slobodan Marko Beroš

**Evaluation of Fatigue Behavior of SAE 9254 Steel Suspension
Springs Manufactured by Two Different Processes:
Hot and Cold Winding** 91
Carolina Sayuri Hattori, Antonio Augusto Couto, Jan Vatavuk,
Nelson Batista de Lima and Danieli Aparecida Pereira Reis

**Yield Criteria for Incompressible Materials
in the Shear Stress Space**.................................. 107
Vladimir A. Kolupaev, Alexandre Bolchoun and Holm Altenbach

**The Optimum Design of Laminated Slender Beams with Complex
Curvature Using a Genetic Algorithm** 121
Jun Hwan Jang and Jae Hoon Kim

**A Finite Element Approach for the Vibration
of Single-Walled Carbon Nanotubes**......................... 139
Seyyed Mohammad Hasheminia and Jalil Rezaeepazhand

Characteristics of Welded Thin Sheet AZ31 Magnesium Alloy 147
Mahadzir Ishak, Kazuhiko Yamasaki and Katsuhiro Maekawa

Localization of Rotating Sound Sources Using Time Domain Beamforming Code 161
Christian Maier, Wolfram Pannert and Winfried Waidmann

Mathematical Modelling of the Physical Phenomena in the Interelectrode Gap of the EDM Process by Means of Cellular Automata and Field Distribution Equations 169
Andrzej Golabczak, Andrzej Konstantynowicz and Marcin Golabczak

Free Vibration Analysis of Clamped-Free Composite Elliptical Shell with a Plate Supported by Two Aluminum Bars 185
Levent Kocer, Ismail Demirci and Mehmet Yetmez

Vibration Analysis of Carbon Fiber T-Plates with Different Damage Patterns.. 195
Ismail Demirci, Levent Kocer and Mehmet Yetmez

Mechanical Characteristics of AA5083: AA6013 Weldment Joined With AlSi12 and AlSi5 Wires 205
Mehmet Ayvaz and Hakan Cetinel

Numeric Simulation of the Penetration of 7.62 mm Armour Piercing Projectile into Ceramic/Composite Armour............... 219
Ömer Eksik, Levent Turhan, Enver Yalçın and Volkan Günay

In-situ TEM Observation of Deformations in a Single Crystal Sapphire During Nanoindentation 229
Fathi ElFallagh, Aiden Lockwood and Beverley Inkson

The Effect of Nanotube Interaction on the Mechanical Behavior of Carbon Nanotube Filled Nanocomposites 241
Beril Akin and Halit S. Türkmen

An Automatic Process to Identify Features on Boreholes Data by Image Processing Techniques 249
Fabiana Rodrigues Leta, Esteban Clua, Mauro Biondi, Toni Pacheco and Maria do Socorro de Souza

An Optimization Procedure to Estimate the Permittivity of Ferrite-Polymer Composite 263
Ramadan Al-Habashi and Zulkifly Abbas

Neural Model for Prediction of Tires Eigenfrequencies

Zora Jančíková, Pavel Koštial, Dana Bakošová, David Seidl, Jiři David, Jan Valíček and Marta Harničárová

Z. Jančíková (✉) · J. David
Department of Automation and Computer Science in Metallurgy, Faculty of Metallurgy and Materials Engineering, VŠB-Technical University of Ostrava, 17. listopadu 15/2172 70833 Ostrava-Poruba, Czech Republic
e-mail: zora.jancikova@vsb.cz

J. David
e-mail: j.david@vsb.cz

P. Koštial
Department of Material Engineering, Faculty of Metallurgy and Materials Engineering, VŠB-Technical University of Ostrava, 17. listopadu 15/2172 70833 Ostrava-Poruba, Czech Republic
e-mail: pavel.kostial@vsb.cz

D. Bakošová
Department of Physical Engineering of Materials, Faculty of Industrial Technologies, University of Alexander Dubček in Trenčín, I. Krasku 491/30 02001 Púchov, Slovak Republic
e-mail: dana.bakosova@fpt.tnuni.sk

D. Seidl
Department of Computer Science, Faculty of Electrical Engineering and Computer Science, VŠB-Technical University of Ostrava, 17. listopadu 15/2172 70833 Ostrava-Poruba, Czech Republic
e-mail: david.seidl@vsb.cz

J. Valíček
Institute of Physics, Faculty of Mining and Geology, RMTVC, Faculty of Metallurgy and Materials Engineering, VŠB-Technical University of Ostrava, 17. listopadu 15/2172 70833 Ostrava-Poruba, Czech Republic
e-mail: jan.valicek@vsb.cz

J. Valíček
RMTVC, Faculty of Metallurgy and Materials Engineering, VŠB-Technical University of Ostrava, 17. listopadu 15/2172 70833 Ostrava-Poruba, Czech Republic

M. Harničárová
Nanotechnology Centre, VŠB-Technical University of Ostrava, 17. listopadu 15/2172 70833 Ostrava-Poruba, Czech Republic
e-mail: marta.harnicarova@vsb.cz

Abstract The work is devoted to the application of an artificial neural network (ANN) to analyze eigenfrequencies of personal tires of different construction. Experimental measurements of personal tire eigenfrequencies by electronic speckle interferometry (ESPI) are compared with those previewed by ANN. Very good agreement of both data sets is presented.

Keywords Tires · Modal analysis · Neural networks · Speckle interferometry

1 Introduction

Important factors in the product development process are the dimensioning of components, the exact determination of material properties, the usage of new materials and the improvement of finite element (FE) calculations. In all of these areas, better understanding of material and component behavior is required, which certainly is a challenge to experimental measuring methods. Needs for a good tire are low rolling resistance, proper hysteresis losses, new tread design, high wear resistance compound and new tire construction.

Tires are the dominant noise sources in vehicles in typical driving conditions. The tire/road noise emission is never omnidirectional as it is generally assumed when used in road traffic noise calculation models.

The influence of tire-pavement interaction and its influence on noise generation were extensively studied in [1]. Onboard sound intensity (OBSI) measurements were taken to quantify the tire pavement noise source strength as a function of pavement parameters. The OBSI results fell into three pavement groupings based on spectral shape. More than other parameters, these groupings were determined by whether the pavement was porous or not and whether it was new or older. The OBSI results also indicated that single-layer porous pavements were particularly effective at reducing tire pavement noise source strength at frequencies above 1,250 Hz for designs 18–33 mm thick. For a thicker, double-layer porous pavement, source strength reductions extended down to 630 Hz.

In the work [2] were determined and compared the directivity patterns of noise from various passenger car tires rolling on various pavements. The selection of pavements consisted of "normal-noise" and "low-noise" pavements including experimental poroelastic pavements. The influence of speed, pavement and tire on noise emission directivity patterns is presented and discussed in this work.

In the work [3] coupling texture and noise data, collected with RoboTex and OBSI, respectively, is serving to advance the state of the art. This work utilized data collected on over 1,000 unique concrete pavement test sections located throughout North America. The ultimate goal of this work is to identify the fundamental links between texture and noise. In the interim, more relevant phenomenological links are sought that have the potential to be expanded to more fundamental models as more is learned about these complex phenomena.

The paper [4] presents the measurement and analysis of rolling tire vibrations due to road impact excitations, such as from cobbled roads, junctions between concrete road surface plates, railroad crossings. Vibrations of the tire surface due to road impact excitations cause noise radiation in the frequency band typically below 500 Hz. Tire vibration measurements with a laser Doppler vibrometer are performed on a test set-up based on tire-on-tire principle which allows highly repetitive and controllable impact excitation tests under various realistic operating conditions. The influence on the measured velocity of random noise, cross sensitivity and alignment errors is discussed. An operational modal analysis technique is applied on sequential vibration measurements to characterize the dynamic behavior of the rolling tire. Comparison between the operational modal parameters of the rolling tire and the modal parameters of the non-rolling tire allows an assessment of the changes in dynamic behavior due to rolling.

Application of electronic speckle interferometry for measurements of tires eigenfrequencies was described in [5]. Authors studied the large amount of tires with different construction and its influence on the eigenfrequency spectrum.

Neural networks are suitable for modeling of complex systems especially from the reason that their typical property is capability of learning on measured data and capability of generalization. Neural networks are able to appropriately express general properties of data and relations among them and on the contrary to suppress relationships which occur sporadically or they are not sufficiently reliable and strong [6]. The application of neural networks in the material engineering and technology were extensively developed also in [7, 8].

In this chapter we present the application of ANN on prediction of eigenfrequencies of personal tires. Predicted data are compared with those experimentally obtained by ESPI.

2 Experimental Procedures

ESPI records the surface displacement of an object in response to the applied force. ESPI can be used in arrangements where fringes will represent lines of either in-plane or out-of-plane displacement.

The out-of-plane set-up can be briefly described as follows: A laser light beam is split into two. One of the beams, the object beam, is used to illuminate the object. A video camera is then used to monitor the illuminated object. The other beam, which is called the reference beam, is directed in such a way that it intersects the view line between the object and video camera. At that point, a partial mirror is used to deflect the reference beam into the video camera making it combine with the light reflected off the object. Due to the monochromatic properties of the laser light, the object and reference beam interfere to produces a unique speckle pattern. The speckle pattern is recorded by the video camera and digitised in a computer in a similar to stereography system.

To perform an ESPI inspection process, a speckle image of the unstressed object is first captured and saved in the computer. The object is then stressed, resulting into the object's surface being displaced.

This causes an alteration of the beam path length of the light reflected off the object surface which in turn causes the unique speckle pattern to change. When compared to the originally stored speckle image, the final image containing the familiar zebra-like fringe patterns is produced. The process for out-of-plane ESPI is highlighted by the equation below:

$$d = \frac{n\lambda}{(\cos\alpha + \cos\beta)} \quad (1)$$

where—λ is the wavelength of laser light, d—out of plane displacement of the object due to the applied stress, α—is the angle between the direction of object normal and camera viewing angle, β—is the angle between the direction of object normal and object beam.

The fringe sensitivity of in-plane displacements is of the order of where α is the angle of incidence of the illuminating beams, which ideally should be the same for both beams.

From the equation above it becomes evident that the object displacement magnitude is constant along a fringe contour, but changes between consecutive fringes due to the increase or decrease in the magnitude of n. The experimental procedure of the ESPI in this case is performed as follows. Tires were excited by a loudspeaker in radial and axial directions. First, a reference image is taken, after the specimen vibrates, then the second image is taken, and the reference image is subtracted by the image processing system. If the vibrating frequency is not the resonant frequency, only randomly distributed speckles are displayed and no fringe patterns will be shown. However, if the vibrating frequency is within the scope of the resonant frequency, stationary distinct fringe patterns will be observed. Then the function generator is carefully and slowly turned, the number of fringes will increase and the fringe pattern will become clearer, as the resonant frequency is being approached. The resonant frequencies and corresponding mode shapes can be determined at the same time, using the ESPI optical system presented in Fig. 1. The measurement of eigenfrequencies was realized by an ESPI apparatus described in [5].

3 Results and Discussion

In the first step we will discuss the eigenfrequency patterns appearing after loudspeaker excitation in both radial and axial direction. The images of standing waves are presented in Figs. 2 and 3.

To test the repetition ability of measurements the every tire was measured five times at every frequency. In Figs. 4 and 5 are presented results of measured radial

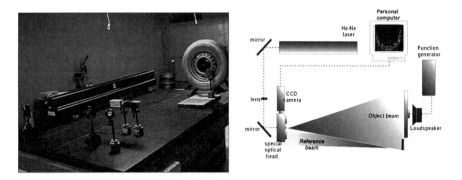

Fig. 1 Real experimental view on ESPI apparatus (*left*) and its schema (*right*)

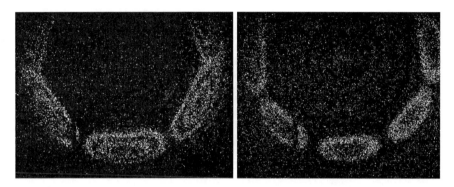

Fig. 2 ESPI mode shape for axial 1.mode, frequency 161 Hz and radial 4.mode, frequency 186 Hz

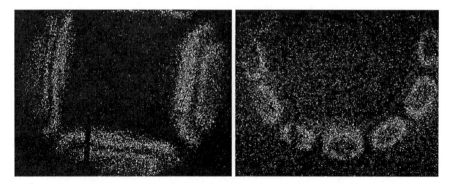

Fig 3 ESPI mode shape for axial 2.mode, frequency 108 Hz and radial 6.mode, frequency 245 Hz

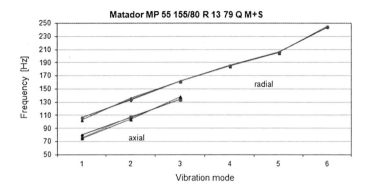

Fig. 4 Reproducibility for different tires of the same dimensions, type 1

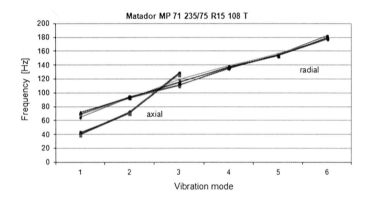

Fig. 5 Reproducibility for different tires of the same dimensions, type 2

and axial eigenfrequencies for different tires of the same producer and the same dimensions. It is possible to see that the dynamic response of different tires correlates very good.

To obtain the qualitative evaluation of the mode structure we use the FE method to model the tire vibrations in the MARC software environment. The Mooney constants were used as material parameters and rubber plasticity model for large deformation was applied. We started the simulation on the model presented in Fig. 6. The simulation was realized under the following:

- Material properties are defined by Mooney—constants.
- We considered the simulated tire without internal reinforcement (cords).
- The rubber plasticity procedure with large deformations was used.
- For the calculation, we have used the Marc Designer software.

Both experimental and simulation results are in very good qualitative harmony. Nevertheless, for quantitative match of the experiment and simulation results, it is

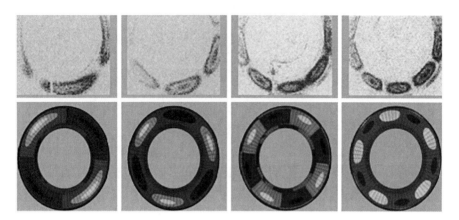

Fig. 6 Picture of measured ESPI loops (*up*) and modeled by FEM (*down*)

Table 1 Frequency versus chosen modes of three different tires producers labelled as C, D, M of the same tire parameters 205/55 R16 91 W

Mode	C (Hz)	D (Hz)	M (Hz)
A1	64	65	81
A2	101	100	96
A3	169	180	175
R1	90	90	66
R2	120	125	122
R3	148	131	140

necessary to know the details of the tire material composition and construction, which is very difficult to obtain because producers usually do not disclose this information. The shape of exited and simulated modes is presented in Fig. 2.

Table 1 presents results of eigenfrequencies *(axial modes A1, A2, A3, radial modes R1, R2, R3)* of three different producers for tires of the same dimensions. The same results are presented in graphical form in Fig. 7. It is clearly seen differences in the tire cavities technological realization (reflected by different eigenfrequencies) which have the same volume in all cases under investigation. Nevertheless differences in eigenfrequencies are apparent in distinguished cases.

In practice, the problems solved by neural networks are most often realized by means of a computer program. Each program unit dealing with the functioning of a neural network can be divided into several concurring parts. These parts correspond to the procedures that are used when working with neural networks. It is a synthesis of the neural network itself, the neural network determination, followed by the actual state when an algorithm solving the problem is designed thanks to the outcomes of the previous operations. The aim is to create such an algorithm that will best describe the modelled system. This algorithm is mostly generated as a

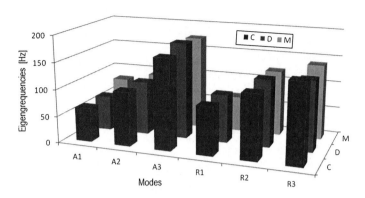

Fig. 7 Graphical output of data presented in Table 1

function in a concrete programming language, which has the required number of input and output parameters.

Of course, there are many implementations of neural networks. The use of many of them is restricted by a license, which means that the use of such software is possible only once you have purchased the appropriate license. However, there are also some freely distributed implementations of neural networks. Freely distributed library of functions necessary for work with FANN neural networks is one of them. The library makes it easy to program the control algorithms using neural networks. The advantage is that the library is intended for many programming languages. The implementation is available in C, C + + , Java, PHP, Matlab languages as well as many others.

The work with FANN library is divided into three parts. In the first part, the programmer must design the network architecture on the basis of his/her experience. It includes the number of input and output neurons, the number of layers and the transition functions of the individual neurons.

A set of test data is applied to the network designed in such a way during the second part. The algorithm uses this designed architecture to perform the process of network learning. The learning is terminated if outputs of sufficient quality are achieved, or if the maximum number of iterations was exceeded. The output of the program is a file that describes the entire network structure with the parameters of the individual neurons. The file describing the newly created neural network is then used to create a program that makes use of the neural network to simulate the behaviour of the system.

There are many parameters influencing what results will be achieved when using the neural network algorithms. The used implementation definitely represents one of the major ones. It can be expected that using commercial solutions will achieve better results, even though it may not always be the case.

Another parameter having absolutely crucial impact on the quality of the entire solution of algorithms using neural networks is the amount of the input data, and especially their homogeneous distribution throughout the entire area in question.

Table 2 Selected measured values

Matador tire labeling	Width (mm)	Height/width ratio (%)	Rim diameter (inches)
MP 59 205/55 R16 91 H	205	55	16
MP 42 225/55 R1695 W	225	55	16
MP 42 185/60 R14 82 T	185	60	14
MP 15 185/6 R14 82H	185	60	14

Matador tire labeling	Load index (kg)	Speed index (km/h)	Radial frequency (Hz)
MP 59 205/55 R16 91 H	91	210	93
MP 42 225/55 R1695 W	95	210	81
MP 42 185/60 R14 82 T	82	190	95
MP 15 185/6 R14 82H	82	210	95

The more data there will be available during the learning stage of the network, the better the results achieved in the prediction stage. When large numbers of data are used, high precision of the measured data is not of crucial importance, as the algorithms of neural networks are to some extent able to eliminate any eventual errors arising during the measurement.

The acquired database for prediction of noise contained 83 cases. Due to the absence of values in case of some variables, the database had been modified and included only 69 cases. These facts were subsequently adjusted to a form suitable for the application of neural network. The whole database was divided into the data to be used for the network learning (training and validation set) and the data to be later used to check the prediction accuracy, i.e. the generalization ability of the neural network (test set). Artificial neural networks were subsequently designed and trained on the basis of the adjusted data. The best prediction results were achieved by a three-layer perceptron neural network of 5-11-2 topology (5 input neurons, 11 in the hidden layer and 2 outputs). Comparative criteria for frequency prediction:

- sum of squares of residues SSE = 1265.871
- mean square error RMS = 4.314
- index of determination R^2 = 0.976

Table 2 shows the tires randomly selected from the test data set, their labelling, parameters and measured values of radial frequency and amplitude. The values of radial frequency and amplitude were predicted for identical tires by the neural network, as indicated in Table 3.

The complex analysis of measured and predicted data shows that the frequency of radial oscillations will increase with the width of the tire and the rim diameter. The dependence of the amplitude of radial oscillations is considerably more complicated. The lowest values can be achieved with increasing width of the tire and decreasing rim diameter. These results were evaluated for a tire with profile number 55, load index 91 and speed index 210.

Table 3 Selected measured values

Matador tire labeling	Width (mm)	Height/width ratio (%)	Rim diameter (inches)
MP 59 205/55 R16 91 H	205	55	16
MP 42 225/55 R1695 W	225	55	16
MP 42 185/60 R14 82 T	185	60	14
MP 15 185/6 R14 82H	185	60	14

Matador tire labeling	Load index (kg)	Speed index (km/h)	Radial frequency (Hz)
MP 59 205/55 R16 91 H	91	210	88.9
MP 42 225/55 R1695 W	95	270	76.5
MP 42 185/60 R14 82 T	82	190	95.6
MP 15 185/6 R14 82H	82	210	94.8

4 Conclusions

The functional neuronal model was created for prediction of tires eigenfrequencies. The model is capable of providing satisfactory prediction of frequency of tires with a mean square error RMS = 4.314 Hz. The presented results and the experience acquired with applications of neural networks in material research show that their utilization in this area is very promising. Neural networks appear to be the functional approximations of relationships between various process data, especially in cases when modeling real systems, which are characterized by a high degree of nonlinearity, considerable complexity and difficulty of formal mathematical description. More accurate results of predictions of various material parameters using neural networks are the result of the fact that the application of neural networks can find links between process parameters which, when conventional methods are used, cannot be traced due to their mutual interactions and a considerable quantity and momentum, and the time-consuming character resulting from those facts.

Acknowledgments This chapter was created in the framework of the project MPO TIP FR-TI1/319. This research was financed also by following projects: MPO TIP FR-TI3/818, InterDV, CZ.1.07/2.2.00/15.0132 and IT4Innovations Centre of Excellence project, reg. no. CZ.1.05/1.1.00/02.0070.

References

1. Donavan, P.R.: Tire noise generation and propagation over porous and nonporous asphalt pavements. Transport Res. Rec. **2233**, 135–144 (2011)
2. Ejsmont, J.A., Mioduszewski, P.: Certification of vehicles used for tire/road noise evaluation by CPX method. Noise Control Eng. J. **57**, 121–128 (2009)
3. Rasmussen, R.: Measuring and modeling tire-pavement noise on various concrete pavement textures. Noise Control Eng. J. **57**, 139–147 (2009)

4. Kindt, P., Sas, P., Desmet, W.: Measurement and analysis of rolling tire vibrations. Opt. Laser Eng. **47**, 443–453 (2009)
5. Kováč, I., Krmela, J., Bakošová, D.: Detection of input material parameters for FEM models of tires. Hutnické listy. **LXIV**, 73–78 (2011)
6. Jančíková, Z.: Artificial neural networks in material engineering. GEP ARTS, Ostrava (2006)
7. Heger, M., Špička, I., Bogar, M., Stráňavová, M., Franz, J.: Simulation of technological processes using hybrid technique exploring mathematical-physical models and artificial neural networks. In: METAL 2011: 20th Anniversary International Conference on Metallurgy and Materials. Ostrava, Tanger (2011)
8. Špička, I., Heger, M., Zimný, O., Červinka, M.: Industrial control systems and data mining. In: Metal 2011: 20th Anniversary International Conference on Metallurgy and Materials. Ostrava, Tanger (2011)

Effect of Steady Ampoule Rotation on Radial Dopant Segregation in Vertical Bridgman Growth

Nouri Sabrina, Benzeghiba Mohamed and Ghezal Abdrrahmane

Abstract For vertical Bridgman growth of metallic binary alloys in a closed ampoule where the flow and dopant transport are not influenced by the upper free surface, we show computationally that the steady rotation about the ampoule axis strongly affects the flow and solute distribution during growth process when the buoyancy convection in the melt is well established. The extended Darcy model, which includes the time derivative and Coriolis terms, has been employed in the momentum equation. The problem is governed by the Navier–Stokes, energy and solute conservation laws. The system obtained has been discretized by the control volume method and resolved by the Patankar's algorithm SIMPLER. A computer code was developed and validated with the previous work. It found that the forced convection introduced by ACRT dramatically changes the flow in the melt and solute distribution at the interface. The effects of Reynolds and Grashof numbers are presented and analyzed.

Keywords Numerical simulation · Unidirectional solidification · Porous media · ACRT · Segregation

---

N. Sabrina (✉) · G. Abdrrahmane
LMFTA, USTHB, BP32 Al Alia, Bab Ezzouar 16111 Algiers, Algeria
e-mail: nouri290676@yahoo.fr

G. Abdrrahmane
e-mail: abdghezal@yahoo.fr

B. Mohamed
IFP Drilling and Completion Department, Pau, France
e-mail: m_benzeghiba@yahoo.fr

1 Introduction

The Bridgman technique is a simple and useful process in growing high quality single crystals [1–3]. However, the heat flow and segregation in the process strongly depend on ampoule orientation and heating uniformity. To minimize the unstable flow and convection, the ampoule is usually aligned with the gravity orientation and the melt sits upon the growth interface, this is the so-called vertical Bridgman (VB) configuration. However, perfect alignment and uniform heating are hard to obtain in practice. To improve the quality of the product obtained, we need to damp the buoyancy convection sufficiently to reach the diffusion-controlled limit [4]. The most well known approach is the use of magnetic damping [5–7]. The flow intensity (ψ_{max}) decreases with the increasing magnetic field strength (B or in terms of Hartmann number Ha); ψ_{max} proportional to Ha^{-2} [8, 9]. However, the hardware requirement to provide a sufficient magnetic damping (usually in the order of 0.5 T) is expensive. An alternative way to suppress the flow, maybe other mechanisms like vibration [10–12] as well is the use of steady ampoule rotation. As discussed by [13, 14], a moderate rotation rate may affect the buoyancy flow significantly for the axisymmetric configuration. In addition, its damping effect is similar to that due to the magnetic field but less effective; ψ_{max} proportional to Ω^{-1} [15], where Ω is the rotation speed.

In the present study, an axisymmetric numerical simulation is conducted to investigate the effects of ampoule rotation on axisymmetric fluid flow and segregation under perfect growth conditions. To further illustrate the feasibility of using ampoule rotation for flow damping, different rotation rates which are represented by the Reynolds numbers are considered and the reduce of buoyancy convection situations, represented by the Rayleigh number, and chemical radial segregations are discussed. In the next section, the model and its numerical simulation are briefly described. Section 3 is devoted to the results and discussion. Finally, we tried to summarize the main physical conclusions of this study.

2 Model Description and Mathematical Formulation

A generic Bridgman crystal growth system is illustrated in Fig. 1. However, the growth dynamics is also considered here. The furnace is described by an effective heating profile $T_a(z, t)$ which is assumed linear. For unsteady state calculations, the thermal profile is unsteady and the ampoule is moved downward at V_g speed.

$$T_a(z,t) = -\frac{(T_h - T_c)}{L_g} \times V_g \times t + T_h \quad (1)$$

where T_h and T_c are respectively the temperatures of the hot and cold zones. L_g is the length of the temperature profile zone.

Effect of Steady Ampoule Rotation

Fig. 1 Schematic of vertical Bridgman growth (VG)

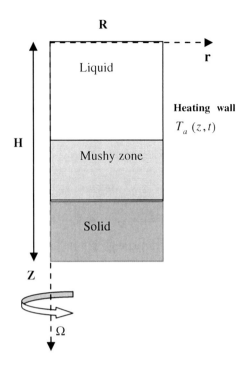

The system is assumed to be axisymmetric and the initial dopant distribution in the melt is assumed uniform at C_0. The flow and solute distribution are represented in a cylindrical coordinate system (r, z). The melt is assumed incompressible and Newtonian, while the flow is laminar. The Boussinesq approximation is also adopted for buoyancy forces modelisation. The stream function ψ is defined in terms of radial (u) and axial (v) velocities as

$$u = -\frac{\partial \psi}{r \partial z}, \quad v = \frac{\partial \psi}{r \partial r} \tag{2}$$

The unsteady-state governing equations for the velocity field (u, v, w), temperature (T) and dopant concentration (C) in the conservative-law can be expressed by:

Continuity equation

$$\frac{1}{r}\frac{\partial (ru)}{\partial r} + \frac{\partial v}{\partial z} = 0 \tag{3}$$

Momentum equation in radial direction

$$\rho\left(\frac{\partial u}{\partial t} + u\frac{\partial u}{\partial r} + v\frac{\partial u}{\partial z}\right) - \boxed{\rho\frac{w^2}{r}} = -\frac{\partial p}{\partial r} + \mu\left(\frac{1}{r}\frac{\partial}{\partial r}\left(r\frac{\partial u}{\partial r}\right) + \frac{\partial}{\partial z}\left(\frac{\partial u}{\partial z}\right) - \frac{u}{r^2}\right) - \underbrace{\frac{\varepsilon u}{l}} \tag{4}$$

The term box represents the centrifugal force, the term (I) represents the flow in the mushy zone. In the problem of metallic alloys phase change, the mushy zone is treated as a porous medium of porosity ζ, see Voller V et Prakash C [16]. According to Darcy's law, ε is derived from Carman-Kozeny's equation [17].

$$\varepsilon = -C(1-\zeta)^2 \Big/ (\zeta^3 + q) \tag{5}$$

with

$$\zeta = 1 - f_S \tag{6}$$

where f_S is the solidification fraction. The constant C depends on the mushy zone's morphology. The constant q is introduced to avoid division by zero.

Momentum equation in axial direction

$$\rho\left(\frac{\partial v}{\partial t} + u\frac{\partial v}{\partial r} + v\frac{\partial v}{\partial z}\right) = -\frac{\partial p}{\partial z} + \mu\left(\frac{1}{r}\frac{\partial}{\partial r}\left(r\frac{\partial v}{\partial r}\right) + \frac{\partial}{\partial z}\left(\frac{\partial v}{\partial z}\right)\right) + \rho g - \varepsilon v \tag{7}$$

Momentum equation in azimuthal direction

$$\rho\left[\frac{\partial w}{\partial t} + v\frac{\partial w}{\partial z} + u\frac{\partial w}{\partial r}\right] + \underbrace{\rho\frac{v.w}{r}}_{II} = \mu_F\left[\frac{\partial}{r\partial r}\left(r\frac{\partial w}{\partial r}\right) - \frac{w}{r^2} + \frac{\partial^2 w}{\partial z^2}\right] - \varepsilon w \tag{8}$$

where *II* represents the coriolis force.

Energy equation

$$\rho C_P\left(\frac{\partial T}{\partial t} + u\frac{\partial T}{\partial r} + v\frac{\partial T}{\partial z}\right) = k\left(\frac{1}{r}\frac{\partial}{\partial r}\left(r\frac{\partial T}{\partial r}\right) + \frac{\partial}{\partial z}\left(\frac{\partial T}{\partial z}\right)\right) - \underbrace{\left[\frac{\partial \rho \Delta H}{\partial t}\right]}_{III} \tag{9}$$

where III expresses the amount of heat released during phase change.

Solute equation

According to Timchenko et al. [18], in solidification problems for a single domain model, the solute transport is governed by the following equation

$$\frac{\partial C}{\partial t} + u\frac{\partial C}{\partial r} + v\frac{\partial C}{\partial z} = D\left[\frac{1}{r}\frac{\partial}{\partial r}\left(r\frac{\partial C}{\partial r}\right) + \frac{\partial}{\partial z}\left(\frac{\partial C}{\partial z}\right)\right] \tag{10}$$

where

$$C = f_s C_s + f_l C_l \tag{11}$$

$$f_s + f_l = 1 \tag{12}$$

$$C_s = k_p C_l \tag{13}$$

f_s and f_l are the solid and liquide fractions. C_s and C_l are solid and liquid concentrations. According to Scheil-Gulliver hypothesis:

$$\partial C_s/\partial t = 0 \tag{14}$$

Thus, Eq. (11) can be partially derived as follows

$$\frac{\partial C}{\partial t} = \underbrace{\left[(1-f_s)\frac{\partial C_l}{\partial t}\right]}_{I} + \underbrace{\left(\frac{\partial f_s}{\partial t}(k_p-1)C_l\right)}_{II} \tag{15}$$

The first term I can be expressed as the change of the solute in the liquid. The second one II defines the rejection of solute in the mushy zone.

The latest version of the solute conservation can be written as follows

$$(1-f_s)\frac{\partial C_l}{\partial t} + u\frac{\partial C_l}{\partial r} + v\frac{\partial C_l}{\partial z} = D\left[\frac{1}{r}\frac{\partial}{\partial r}\left(r\frac{\partial C_l}{\partial r}\right) + \frac{\partial}{\partial z}\left(\frac{\partial C_l}{\partial z}\right)\right] + \frac{\partial f_s}{\partial t}(k_p-1)C_l \tag{16}$$

Solid fraction calculation

In metallurgical solidification of binary alloy, the function $f_s(T)$ will depend on the nature of the solute distribution and the associated phase change equilibrium diagram. In the present work a simple linear form for the local solid fraction $f_s(T)$ is chosen:

$$f_s(T) = \begin{cases} 0 & T \geq T_L \\ \frac{T_L-T}{T_L-T_s} & T_L \succ T \geq T_s \\ 1 & T \prec T_s \end{cases} \tag{17}$$

where T_L is the liquidus temperature at which solid formation commences and T_s is the solidus temperature

Non–dimensional equations

The reduced variables are given by the basic reference variables as follow:

$$t^* = u_0 t/R \quad r_* = r/R \quad Ste = Cp\,\Delta T/\Delta H \quad u_* = u/u_0$$
$$V_* = V/u_0 \quad p^* = p/\rho_0 u_0^2 \quad T^* = (T-T_c)/(T_h-T_c) \quad C^* = C/C_0$$

where R is the radius of the cavity, $(u_0 = \Omega R)$ is the reference speed based on the rotational speed Ω, $(\Delta T = T_h - T_c)$ is the temperature difference between hot and cold zones, C_0 is the reference concentration and $\rho_0.u_0^2$ is the reference pressure.

The system of equations and the boundary conditions which govern the problem are written in the following dimensionless form

$$\frac{1}{r^*}\frac{\partial(r^*u^*)}{\partial r^*}+\frac{\partial v^*}{\partial z^*}=0$$

$$\frac{\partial u^*}{\partial t^*}+u^*\frac{\partial u^*}{\partial r^*}+v^*\frac{\partial u^*}{\partial z^*}-\frac{w^{*2}}{r}=-\frac{\partial p^*}{\partial r^*}+\frac{1}{Re}\left(\frac{1}{r^*}\frac{\partial}{\partial r^*}\left(r^*\frac{\partial u^*}{\partial r}\right)+\frac{\partial}{\partial z^*}\left(\frac{\partial u^*}{\partial z^*}\right)-\frac{u^*}{r^{*2}}\right)$$

$$-\frac{\eta\varepsilon}{Re}u^*\left(\frac{\partial v^*}{\partial t^*}+u^*\frac{\partial v^*}{\partial r^*}+v^*\frac{\partial v^*}{\partial z^*}\right)=-\frac{\partial p^*}{\partial z^*}+\frac{1}{Re}\left(\frac{1}{r^*}\frac{\partial}{\partial r^*}\left(r^*\frac{\partial v}{\partial r^*}\right)+\frac{\partial}{\partial z^*}\left(\frac{\partial v^*}{\partial z^*}\right)\right)$$

$$-\frac{\eta\varepsilon}{Re}v^*+\left(1-\frac{Ra\,Pr}{Re^2}((T^*-1)+N(C^*-1))\right)\cdot\left[\frac{\partial w^*}{\partial t}+v\frac{\partial w^*}{\partial z}+u\frac{\partial w^*}{\partial r}\right]$$

$$+\left(1-\frac{Ra\,Pr}{Re^2}((T^*-1)+N(C^*-1))\right)\frac{v^*.w^*}{r}=\frac{1}{Re}\left[\frac{1}{r}\frac{\partial}{\partial r}\left(r\frac{\partial w^*}{\partial r}\right)-\frac{w^*}{r^2}+\frac{\partial^2 w^*}{\partial z^2}\right]$$

$$-\frac{\eta\varepsilon}{Re}w^*\left(\frac{\partial T^*}{\partial t^*}+u^*\frac{\partial T^*}{\partial r^*}+v^*\frac{\partial T^*}{\partial z^*}\right)=\frac{1}{Pr\,Re}\left(\frac{1}{r^*}\frac{\partial}{\partial r^*}\left(r^*\frac{\partial T^*}{\partial r^*}\right)+\frac{\partial}{\partial z^*}\left(\frac{\partial T^*}{\partial z^*}\right)\right)$$

$$+\frac{1}{Ste}\left[\frac{\partial f^s}{\partial t^*}\right]$$

$$\left((1-f_s)\frac{\partial C_l^*}{\partial t^*}+u\frac{\partial C_l^*}{\partial r^*}+v\frac{\partial C_l^*}{\partial z^*}\right)=\frac{Sc}{Re}\left[\frac{1}{r^*}\frac{\partial}{\partial r^*}\left(r^*\frac{\partial C_l^*}{\partial r^*}\right)+\frac{\partial}{\partial z^*}\left(\frac{\partial C_l^*}{\partial z^*}\right)\right]$$

$$+(k_p-1)C_l^*\frac{\partial(f_s)}{\partial t^*} \tag{18}$$

The no slip boundary condition on velocity is applied:
$0\leq z^*\leq\frac{H}{R},\quad 0\leq r^*\leq 1$:

$$u^*=v^*=0 \tag{19}$$

A no—flux condition is imposed at the top and the bottom of the cavity while the side wall is exposed to a uniform temperature gradient
$0\leq r^*\leq 1$:

$$\left.\frac{\partial T^*}{\partial r^*}\right|_{z^*=0}=\left.\frac{\partial T^*}{\partial r^*}\right|_{z^*=H/R}=0 \tag{20}$$

$0\leq z^*\leq\frac{H}{R}$:

$$T^*(z^*)\big|_{r^*=1}=-\left(V_g^*/L_g^*\right)\times t^*+1 \tag{21}$$

The cavity walls are impermeable to any mass transfer
$0\leq r^*\leq 1$:

$$\left.\frac{\partial C^*}{\partial r}\right|_{z^*=0}=\left.\frac{\partial C^*}{\partial r^*}\right|_{z^*=H}=0 \tag{23}$$

$0\leq z^*\leq\frac{H}{R}$:

$$\left.\frac{\partial C^*}{\partial z^*}\right|_{r^*=R} = 0 \tag{24}$$

The axis of the cylinder is taken as line of symmetry for all field variables

$$\left.\frac{\partial u^*}{\partial z^*}\right|_{r=0} = \left.\frac{\partial v^*}{\partial z^*}\right|_{r=0} = \left.\frac{\partial T^*}{\partial z^*}\right|_{r=0} = \left.\frac{\partial C^*}{\partial z^*}\right|_{r=0} = 0 \tag{25}$$

Pr is the Prandtl number $(Pr \equiv \rho C_p \nu/\lambda)$, Sc the Schmidt number $(Sc \equiv \nu/D)$. ν is the kinematic viscosity and D the dopant diffusivity in the melt. The Grashof number Gr and the buoyancy ration N in the source terms of the momentum equation are defined as follows:

$$Ra \equiv g\rho C_p \beta_T \Delta T\, R^3/\nu\lambda, \quad N \equiv \beta_C \Delta C/\beta_T \Delta T$$

where g is the gravitational acceleration, $\Delta T = T_h - T_c$, β_T and β_C the thermal and solutal expansion coefficients, respectively. Other, the centrifugal or rotation-driven flows are generated by the rotation of the ampoule. The flow pattern depends on rotation rates, container radius and melts properties. The characteristic rotational Reynolds number is the ratio between inertia forces and viscous forces:

$$\mathrm{Re} = \Omega R^2/\nu$$

where Ω is a characteristic rotation rate of the crystal.

The Stefan number $Ste \equiv \Delta H/(C_p \Delta T)$ scales the heat fusion ΔH released during solidification to the sensible heat in the melt.

As in our previous work [19], dopant rejection at the interface is accounted by using the segregation coefficient k_p and dopant diffusion in the solid is neglected. Compared to the nonrotating case, we need to account for an additional velocity component, w in the azimuthally direction. The only twitch will be on the dynamic boundary conditions where $w^* = 0$ on the inner's wall ampoule and $w^* = 1$ on the lateral surface.

3 Numerical Procedure

The obtained coupled system of equations, in one domain approach, includes the convection in the liquid part and mushy zone as well as the diffusion in the solid part and the solute distribution between the three phases (liquid, mushy zone and solid). This system is solved by the finite volume method using a primitive variables (pressure - velocity) formulation; see Patankar [20].

The principle of this method is based in integrating the equation considered on the control volume and to evaluate the various variables not located on the calculation grid by adequate interpolations. This fundamental property will allow the description of the conservation properties of local and total flows and fluxes.

In order to optimize the computing time, we choose the line by line algorithm in coupling with the correction per blocks method. The algorithm SIMPLER is selected to deal with pressure - velocity coupling. A power law model is adopted to fit with the interpolation of diffusive and convective terms. The enthalpy-porosity formulation for problems involving phase change was successfully used for directional solidification in unsteady state. Most of the simulation used the grid of $N_r = 30$ and $N_z = 90$ and the time step of $\delta t = 0.5$ according to a discretization dependency study.

4 Results and Discussion

We present our results for an annular cavity of aspect ratio $A = 3$. In our numerical simulations, the melt is initially at liquid state defined by the Prandtl number equal to 0.01. The rejected solute at the liquid-mushy zone interface is defined by the Schmidt number of Sc = 10 and by segregation coefficient $k_p = 0.1$. To well follow the vertical solidification evolution and the process of the partition of solute in the mushy zone, $\left(V_g^*/L_g^*\right) = 8 \times 10^{-3}$ is chosen. We have computed thermally driven flow, solute distribution and radial segregation δC as a function of rotation rate. The radial segregations are calculated by the following relationship:

$$\delta C = (C_{\max} - C_{\min})/C_{av}$$

Where $C_{\max}$, $C_{\min}$ and C_{av} are, respectively, the maximum, minimum and radially average interfacial concentration. To evaluate the simultaneously effects of Grashof and Reynolds numbers on the flow and solute distribution, we have chosen a typical case where the thermal and solutal buoyancy forces are opposite (in the case which the solvent is heavier than the solute). The Grashof number depends on $(T_h - T_c)$, in our case $Ra \prec 10^7$. This values interval guarantees the laminar regime (El Ganaoui et al. [21]). The Reynolds number, which represents the applied coriolis force intensity, varies from 0.1 to 10. The values of the two last physical sizes are independent.

The physical properties and some input parameters are listed in Tables below for reference (Tables 1, 2).

a. **Effect of Rayleigh Number**

For the cases without ampoule rotation, the unsteady state results at different convection levels are shown in Fig. 2. To better visualize the convection's effect, these results were taken at $t^* = 30$ where growth has been stared ($f_s = 0.13$). In each plot, the left-hand side represents the iso-concentration lines, while the right-hand side represents the solidification fraction function contours. Between the two plots, stream function contours.

In this case, we see the appearance of the mushy zone ($0 \leq f_s \leq 1$) where the flow is almost absent. This is demonstrated by the fact that the solute's distribution

Table 1 Experimental input parameters for the vertical Bridgman system [22]

Description	Symbol	Value
Crystal radius (m)	R	5×10^{-3}
Crystal length (cm)	H	70×10^{-3}
Pulling rate (m/s)	V_g	8×10^{-7}
Initial dimensionless concentration	$C = c/C_0$	1
Temperature difference (K)	$\Delta T = T_h - T_c$	$10–10^3$
Gradient temperature length (m)	L_g	0.138

Table 2 Thermo physical property data used in analysis [22]

Quantity	Symbol	Value
Density (kg/m³)	ρ	5.5×10^3
Kinematic viscosity (m²/s)	ν	1.3×10^{-7}
Thermal conductivity (W/°C m)	λ	27.8
Specific heat (J/ °C.Kg)	c_p	390
Thermal expansion coefficient of (C^{-1})	β_T	5×10^{-4}
Solutal expansion coefficient of (C^{-1})	β_S(mole fraction)$^{-1}$	-0.3
Diffusion coefficient (m²/s)	D	1.3×10^{-8}
Partition coefficient	k_p	0.1
Latent heat (J/kg)	ΔH	460×10^3

in the mushy zone is almost diffusive. Above the mushy zone, the flow becomes very important and increases as Rayleigh number increases, which results in poor solute's distribution at the interface; liquid-mushy zone.

b. Effect of Reynolds Number

The first observation we can make is that the phase change's phenomenon is a little late in the case of rotation. We have had $f_s = 0.13$ at $t^* = 49$. To demonstrate the rotation's effect, we have chosen the case where the convection is well pronounced. By comparing the figures below, we see that the rotation creates a flow in the mushy zone which was almost absent in the case without rotation (Fig. 4a). This flow created reviewing the solute's distribution at the interface and reduces the radial segregation (Fig. 3). As the generated flow by the coriolis force is proportional to Re^{-1}, when the Reynolds number increases, the génerated flow in the mushy zone decreases. For Re $\succ$ 1, the convection in the melt becomes weak and the radial segregation becomes very important at the interface. This is due to the fact that the forced convection created by rotation and the natural convection generated by gradient temperature are in confrontation. To ensure well redistribution of solute at the interface, the flow generated by rotation must be sufficiently greater than that generated by the temperature profile (Fig. 4).

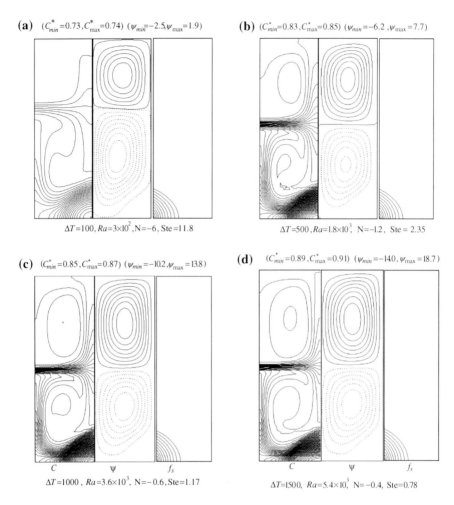

Fig. 2 Calculated flow patterns and dopant field for $f_s = 0.13$

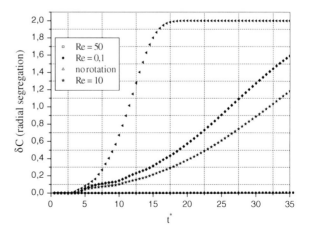

Fig. 3 Effect of rotation rate on radial segregation during solidification

Effect of Steady Ampoule Rotation

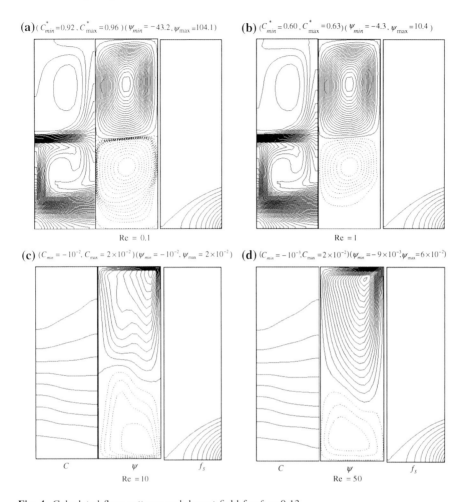

Fig. 4 Calculated flow patterns and dopant field for $f_s = 0.13$

5 Conclusion

The ampoule rotation effect on the flows and dopant segregation in vertical Bridgman crystal growth is investigated numerically. From the calculated results, it is clear that a moderate ampoule rotation speed can significantly affect the flows, solidification fraction and further the dopant mixing. Similar to axially magnetic damping, the Coriolis force due to ampoule rotation can suppress thermal convection. The flow inhibition by ampoule rotation can be regarded as an extension of the Taylor-Proudman theorem for inviscid melt. In summary, the ampoule rotation is believed to have large effects on vertical Bridgman crystal growth.

However, how to better control the melt convection and dopant transport by ampoule rotation still requires further study, mainly, through crystal growth experiments.

References

1. Gault, A., Monberg, E., Clemans, J.: A novel application of the vertical gradient freeze method to the growth of high quality. J. Cryst. Growth **74**, 4–491 (1986)
2. Hoshikawa, K., Nakanishi, H., Kohda, H., Sasaura, M.: Homogeneous increase in oxygen concentration in Czochralski silicon crystals by a cusp magnetic field. J. Cryst. Growth **94**, 635–643 (1989)
3. Monberg, E., Gault, W., Simchock, F., Domingguez, F.: Vertical gradient freeze growth of large diameter, low defect density indium phosphide. J. Cryst. Growth **83**, 160–174 (1987)
4. Jun, L., Bofeng, B.: Thermosolutal convection and solute segregation during the vertical Bridgman growth of Hg1−xCdxTe single crystals. J. Cryst. Growth **311**, 20–38 (2008)
5. Mitric, A., Duffar, Th: Design of vertical Bridgman experiments under alternating magnetic field. J. Cryst. Growth **310**, 1490–1511 (2008)
6. Xi, L., Zhongming, R., Guanghui C., Fautrelle,Y.: Structure, growth characteristic and magnetic properties in directionally solidified Bi-MnBi composite under strong magnetic field. Acta. Materialia. **59**, 6297–6307 (2011)
7. Henry, D., Ben Hadid, H., Kaddecheb, S., Dridi, W.: Macrosegregation and convection in the horizontal Bridgman configuration. J. Cryst. Growth **310**, 1533 (2008)
8. Nouri, S., Benzeghiba, M., Benzaoui, A.: Numerical study of the vertical solidification process. Energy Procedia **6**, 531–540 (2011)
9. Jun, L., Bofeng, B.: J. Cryst. Growth **311**, 38 (2008)
10. Liua, Y., Yub, W., Rouxc, B., Lyubimovad, T., Lana, C.: Chem. Eng. Sci. **61**, 7766 (2006)
11. Liua, Y., Yub, W., Rouxc, B., Lyubimovad, T., Lana, C.: The effect of sensitization on the hydrogen-enhanced fatigue crack growth of two austenitic stainless steels. J. Cryst. Growth **311**, 684 (2009)
12. Zhang, Y., Liu a, Jiang, W., Pan, X., Jina, W., Ai, F.: Segregation control of vertical Bridgman growth of Ga-doped germanium crystals by accelerated crucible rotation: ACRT versus angular vibration. J. Cryst. Growth **311**, 310–5432 (2008)
13. Capper, P., Maxey, C., Butler, C., Grist, M., Price, J.: Bulk growth of cadmium mercury telluride (CMT) using the Bridgman/accelerated crucible rotation technique (ACRT). J. Cryst. Growth **275**, 259–275 (2005)
14. Liu, Y., Roux, B., Lan, C.: Effects of accelerated crucible rotation on segregation and interface morphology for vertical Bridgman crystal growth: Visualization and simulation. J. Cryst. Growth **304**, 236–243 (2007)
15. Lan, C., Liang, M., Chian, J.: Phase field modeling of convective and morphological instability during directional solidification of an alloy. J. Cryst. Growth **295**, 212–340 (2006)
16. Voller, V., Prakash, K.: A fixed grid numerical modeling methodology for convection-diffusion mushy region phase-change problems. Int. J. Heat Mass Transf. **30**, 1709–1719 (1987)
17. Carman, P.: Fluid Flow through Granular Beds. Trans. Inst. Chem. Eng. **15**, 150–166 (1937)
18. Timchenko, V., Chen, P., Leonardi, E., de Vahl, Davis G., Abbaschian, R.: A computational study of transient plane front solidification of alloys in a Bridgman apparatus under microgravity conditions. Int. J. Heat Mass Transf. **43**, 963–980 (2000)
19. Nouri, S., Benzeghiba, M., Benzaoui, A.: Numerical analysis of solute segregation in directional solidification under static magnetic field. Defect Diffus Forum **312**, 253–258 (2011)

20. Patankar, S.: Numerical heat transfer and fluid flow. Mc Graw Hill, New York (1980)
21. El Ganaoui, M., Bontoux, P., Morvan, D., Acad, D.: Localisation d'un front de solidification en interaction avec un bain fondu instationnaire. Sci. Paris **327**, 41–48 (1999)
22. Sampath, R., Zabaras, N.: Inverse thermal design and control of solidification processes in the presence of a strong external magnetic field. Int. J. Numer. Meth. Eng. **50**, 2489–2520 (2001)

Discontinuity Detection in the Vibration Signal of Turning Machines

Joško Šoda, Slobodan Marko Beroš, Ivica Kuzmanić and Igor Vujović

Abstract The chapter deals with the detection of discontinuities in the vibration signal created by a turning machine in soft and hard processes. This chapter presents an experiment with two embedded piezo-electrical sensors on the turning machine tool tip. AISI 4140 steel workpieces are used in the experiments to obtain the analyzed signal. The purpose of the experiment is to identify the nature and position of rapid changes as a measure of compliance with surface roughness. A new algorithm for wavelet selection is developed and the procedure proved the Daubechies wavelet of the 6th order to be the best choice. The proposed algorithm is based on a new technique of energy matching for wavelets. Due to the algorithm's efficiency, it is well suited for real time monitoring. Furthermore, it is possible to build a real time monitoring system based on the digital signal system.

Keywords Vibration signal · Wavelet transform · Tool condition monitoring · Turning machine

---

J. Šoda · I. Kuzmanić (✉) · I. Vujović
Faculty of Marine Studies, University of Split, Zrinsko-Frankopanska 38, 21000 Split, Croatia
e-mail: ikuzman@pfst.hr

J. Šoda
e-mail: jsoda@pfst.hr

I. Vujović
e-mail: ivujovic@pfst.hr

S. M. Beroš
Faculty of Electrical Engineering, Mechanical Engineering and Naval Architecture, University of Split, Ruđera Boškovića bb, 21000 Split, Croatia
e-mail: sberos@fesb.hr

1 Introduction

The impact of cutting conditions on cutting forces and vibration signals [1] produces economic costs, making the monitoring of wear in turning operations by measuring vibration and strain imperative [2]. It is known that wavelets can be used to detect discontinuities on metal surfaces due to the processing of the insert in the turning process [3]. The locations of discontinuities correlate with the damage on the metal surface [4]. Measurement analysis indicates that the Morlet wavelet is suitable for the illustration of the distribution of amplitude and frequency of the turned surface. Mexican hat wavelet is suitable for the detection of significant changes in the vibration signal, indicating surface discontinuities of the insert [3].

Nowadays, there is a growing demand for autonomous systems for monitoring critical parameters (tool condition monitoring, TCM) in industry to improve the efficiency of manufacturing, processes and/or machines. Autonomous or integrated measurement systems are based on the use of sophisticated microprocessor technology, which enables laboratory and industrial measurements, monitoring and classification of critical parameters in real time. These trends are also current in the turning process [5]. Due to the complexity of the turning process, the processing of metal by turning has been studied for over a century. Some aspects of process dynamics are still not sufficiently investigated [1].

The typical processes in turning machine operation are classified by the following characteristics:

- complex and chaotic behavior due to the non-homogeneous structure of the insert,
- sensitivity of process parameters, and
- nonlinear relationship between the parameters of process and knife wear and tear.

In the beginning, the monitoring of the turning process relied on human eyes, ears and nose, which made it very subjective. Process automation was made possible by the development of sensors. Monitoring systems have to meet the following requirements:

- advance error detection in the cutting system and the insert,
- checking and safeguarding the machine process stability,
- keeping the tolerance within parameters, and
- avoiding the malfunctioning of the system.

The goal of every machine processing is to remove as much of the insert's volume as possible in the shortest period possible within the required parameters. In the process, phenomena like chatter occur when for different reasons the blade becomes separated from the insert during turning, making holes in the material at the moment of their reunion. The blade can become separated from the insert for a variety of reasons:

- low-quality material,
- material with holes,
- blunt blade, etc.

Vibrations occur during machine processing. The vibrations make indentations on the processed surface. Due to the complexity of machine processing, there is no direct method with which to predict the result on the insert's surface. Furthermore, there are so many parameters to be considered that it would be highly impractical to take them all into account.

Examples of parameters hard to control or model are:

- process duration changes due to the wear of tools,
- the point of contact between the tool and the insert,
- time and mechanical properties of the system,
- thickness of an insert does not depend solely on vibrations, but also on the pre-existent surface conditions.

These changes have a direct influence on the deviations from the required geometrical shape.

This chapter presents an experiment aiming to identify the nature and location of the discontinuities. The energy of wavelets is used to localize discontinuities in the vibration signal in order to improve on-line diagnostics. The chapter presents an original experiment with two embedded piezo-electrical sensors on the turning machine tool tip.

The chapter is conceived as follows. The theory of wavelets and the engineering problem are presented in the Sect. 2. In the Sect. 3, an algorithm for wavelet selection is introduced. The Sect. 4 deals with experimental setup and fifth with results. The Sect. 5 concludes and summarizes the results and the proposed method.

2 Theory

Wavelet theory is well known and explained in many references, such as [6–9]. It was considered almost revolutionary in different applications. Great advances were made in signal processing and the analysis of 1-D problems [10]. Wavelets were also hugely successful in the suppression of vibration errors in video signals [11].

Continuous wavelet transform (CWT) is defined by:

$$CWT(a,b) = \frac{1}{\sqrt{a}} \int_{-\infty}^{\infty} f(t) \cdot \psi * \left(\frac{t-b}{a}\right) dt = \langle \psi_{a,b}(t), f(t) \rangle \qquad (1)$$

where:

- $a \in R^+$ designates the scaling factor (parameter which directly influence resolution in frequency domain),

- $b \in R$ designates the position (influences time resolution of the signal),
- $f(t)$ the analyzed signal,
- $\psi_{a,b}(t)$ basic functions.

Discrete wavelet transform (DWT) was developed for calculations in discrete-time and computer applications. DWT can be represented as:

$$W(s,\tau) \cong \int x(t) \cdot \psi_{j,k}^*(t) dt = d_{j,k} \qquad (2)$$

and inverse DWT as:

$$x(t) = \frac{1}{c} \sum_{j=-\infty}^{\infty} \sum_{k=-\infty}^{\infty} d_{j,k} \cdot \psi_{j,k}(t) \qquad (3)$$

where the constant, c, equals to:

$$c = \int_{-\infty}^{\infty} \frac{|\psi(\omega)|}{|\omega|} d\omega \qquad (4)$$

under condition that:

$$\int_{-\infty}^{\infty} \psi(t) dt = 0. \qquad (5)$$

In this chapter, only orthogonal wavelet basis are tested. It is necessary to limit the set of wavelet families used, because there is an infinite number of wavelets. It is known from practice that orthogonal wavelets are best suitable for real-time applications, which is the goal in this research.

Definition 2.1 [12] If $\{v_j\}$ is a set of orthonormal basis for V, than any w in V can be written as:

$$w = \sum_{j=1}^{n} \langle w, v_j \rangle v_j = \langle w, v_1 \rangle v_1 + \langle w, v_2 \rangle v_2 + \ldots + \langle w, v_n \rangle v_n \qquad (6)$$

Definition 2.2 [12] Let W be a finite dimensional subspace of a vector space V with the Euclidean inner product. If $\{v_1, v_2, \ldots, v_m\}$ is an orthogonal basis for W, and u is in V, then:

$$P_w u = \sum_{j=1}^{m} \frac{\langle u, v_j \rangle v_j}{\|v_j\|^2} \qquad (7)$$

where $P_w u$ is the orthogonal projection of u on to the subspace W spanned by $\{v_1, v_2, \ldots, v_m\}$.

As can be seen from abovementioned definitions, orthogonal wavelets produce only diagonal matrices, which save the execution time and therefore it makes them suitable for engineering applications.

Examples of the orthogonal wavelets are:

- Daubechies,
- symlets and
- coiflets.

An important property of the orthogonal wavelets is the fact that both the scaling function and the wavelet function have a compact support (finite domain). Another property of the orthogonal wavelets is the number of vanishing moments that are equal to zero. The orthogonal wavelet families are used in both the continuous and the discrete wavelet transform. The orthogonal wavelets are not regular. Daubechies wavelets are asymmetrical. Symlets are almost completely symmetrical. Coiflets have scaling and wavelet functions symmetrical.

One can ask why not to develop a novel wavelet. Although it could be done, it is not necessary, because known wavelets have operated within satisfactory accuracy. The development of the novel wavelet means the design of new approximation and detail filters, which might produce better results, but it is discussible whether it is worth of effort.

An insert processed by the turning machine has a rough surface. The measure of profile roughness is the mean value of distances from the mid-line and is defined by:

$$R_A = \frac{1}{\ell} \int_0^\ell |y| dx \qquad (8)$$

where:

- R_A is the measure of profile roughness,
- ℓ is the insert length,
- y height of the surface point, and
- x surface length element.

In order to monitor and analyze this signal, we need to know something about the expected spectrum.

An example of a spectrum is shown in Fig. 1. Figure 1 shows the expected spectrum for the procedure of the deep cut. It is a 3D graph showing amplitude, frequency and cutting depth. It can be seen that the cutting depth does not vary significantly at higher frequencies, but it varies considerable at the lower frequencies. Therefore, it can be concluded that variation of cutting depth can be expected at the lower frequency bands, which is in the range of the approximation wavelet coefficients.

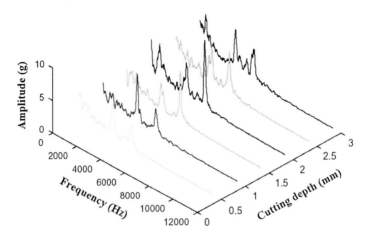

Fig. 1 Expected spectrum for procedures of deep cut [13]

It is important to point out that the location of discontinuities in the vibration signal correlates with the appearance of damage on the insert.

3 Proposed Solution

After turning, the insert undergoes a surface roughness check, as a measure of quality of the turning. The turned inserts are placed into boxes and a certain number of inserts is selected from the box. The surface roughness is measured by randomly choosing a number of points on the insert at which surface roughness is measured.

The proposed method of processing is actually a diagnostic device, which uses wavelets to pinpoint the position of all pre-existent changes or changes occurring on the material during processing, in real time. The surface roughness is no longer a statistical method, because the wavelets are showing us exactly which insert needs to be checked. The proposed method is to choose a wavelet best matching a part of vibration signals. Since vibration signals are always similar, a piece is cut off and error minimized, i.e. square of difference of the wavelet signal spectrum and spectrum of the cut-off vibration signal. In our experience, the obtained results show that db5 (Matlab designation for the Daubechies wavelet with 5 vanishing moments) is the most suitable for this purpose, i.e. it has the smallest square of difference.

Since cost-effectiveness is of essence in the capitalist economy, the "problem" with the soft processing of metals is that the processing takes longer and the blade blunts faster. On the other hand, the positive side of soft processing of metals is that the chatter effect only occurs on defective materials. The "problem" with soft turning is that chatter effects are not clearly visible in the vibration signal, consequently we had to perform some filtering.

The filtering was performed by taking and filtering only the third and fourth levels of the decomposition of signals with wavelets by the application of the hard threshold method, with 2 sigma limits.

The method was used due to the existence of a natural frequency of the blade and the insert at the third and fourth levels of decomposition, so that when the blade is separated from the insert, the change is only registered around the natural frequencies (due to their predominance).

This is the reason why the signal was only reconstructed around s3 (the third level of decomposition) and s4 (the fourth level of decomposition) to obtain the position for the execution of quality control. This cannot be seen at other levels of the decomposition, because in soft processing, since the turning blade is shallowly set and the speed of turning is low, the chatter effect can only occur in low-quality materials. The chatters (separations of the blade from the insert) can only be seen in the environment of resonant frequencies (frequency of the blade), due to their predominance. The rest is lost in the noise of vibration signals.

The story is completely different in case of rough processing of metal. Since the blade cuts the material deeper and the processing is fast, the chatter effect is a frequent occurrence, i.e. the blade can also become separated from the insert due to the speed of turning. The probability that the material will not pass the surface roughness control is higher. The filtering is likewise applied and hard threshold with 2 sigma limits established. All levels of decomposition can be taken into account, because the separations occur at all levels when the blade is separated from the insert. Since the resonant frequencies of the blade and the insert are no longer predominant, chatter spills over the entire spectrum. Besides, if the approximation signal remaining after the vibration signal is decomposed into 5 levels is examined, spikes (chatters) are immediately discernible in the approximation signal during rough processing of metal, because approximation signal actually represents the appearance of the surface of the material after the separation of all details (stripping of the resonant frequency of the blade, etc.).

This is extremely important, because apart from the fact that wavelets reveal the chatter effect, the appearance of the surface after processing can also be discerned from the vibration signal.

Morlet wavelet is well suited to the description of the distribution of amplitude and frequency in surfaces processed by the turning machine [3]. Mexican hat wavelet was proven to be good for the detection of changes in signal causing discontinuities on the insert's surface [3]. In this research we developed an algorithm for the selection of a wavelet best suited to the detection of discontinuities.

Figure 2 shows the algorithm flow.

The wavelet selection is based on estimation for the cost function, which is the difference between the energy of a wavelet and the characteristic part of the vibration signal in time or frequency domain, taking into account the pseudo-frequency of wavelets in the function of the adopted wavelet. It can be written as [4]:

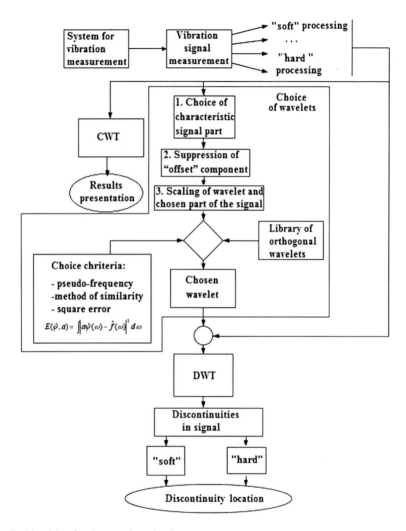

Fig. 2 Algorithm for the wavelet selection

$$\varepsilon^2 = \sum_n \left(\hat{\psi}(\omega) - S(\omega)\right)^2 \qquad (9)$$

The algorithm begins with the selection of the characteristic part of the vibration signal. The second operation in the algorithm is the suppression of the "offset" component, because wavelets are suited for signals without it. In order to make comparisons and decisions the signal's part of interest needs to be normalized. The decision on the wavelet selection is based on pseudo-frequency, the method of similarity and square error. For an arbitrary wavelet from library, the square error is calculated:

$$E(\hat{\psi}, a) = \int \left|a\hat{\psi}(\omega) - \hat{f}(\omega)\right|^2 d\omega \quad (10)$$

The best suited wavelet is the one with minimum square error. For the selected wavelet, discrete wavelet transform (DWT) is performed and discontinuities in the signal are detected.

In the research, Daubechies wavelet of the 6th order (Matlab designation db6) is found to be the best based on the criterion from Eq. (9).

The wavelet pseudo-frequency can be calculated by the following equation:

$$f_{ps} = \frac{f_c}{a\Delta t} \quad (11)$$

where f_{ps} represents the pseudo-frequency of the chosen wavelet, fc represents the natural frequency of the chosen wavelet, a represents scale factor (which is in this research equal to 10), and t represents the sampling time. The sampling time can be obtained by:

$$\Delta t = \frac{1}{f_s} = 22.675 \, [\mu s] \quad (12)$$

4 Experimental Setup

The frequency range of the vibration signal is localized in the audio frequency band [1, 14, 15]. Data acquisition is performed by Sound Blaster Extigy card, manufactured by Creative®, built into the PC. The PC's operating system is Windows XP. The signal is acquired by the application of Cool Edit® version 1.1. The program application packet Matlab® is used in the analysis of the measured vibration signal. The experiment is conducted to identify the nature and location of discontinuity as measure of roughness. Figure 3 shows the scheme of the experiment. Figure 4 shows the actual appearance of the experimental setup. Inserts of AISI 4140 steel are used in the experiments to obtain the analyzed signal (Fig. 4, see material's sample).

The measurement of the tool's natural frequency is performed several times. The average value of the natural frequency is 3.25 [kHz].

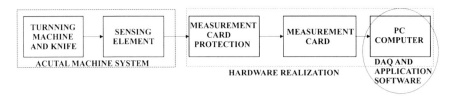

Fig. 3 Scheme—experimental setup

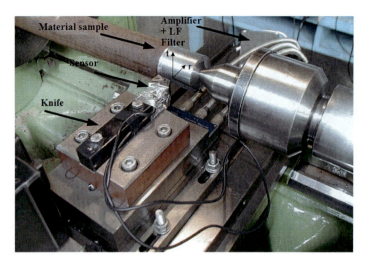

Fig. 4 Photograph—experimental setup

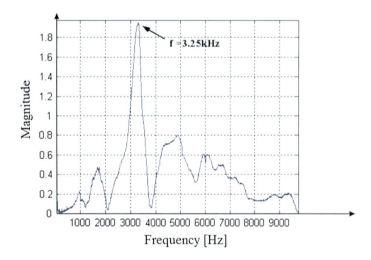

Fig. 5 Tool's natural frequency measurement results

Typical measurement result can be seen in Fig. 5.

It can be seen from Fig. 5 that several considerable frequencies exist. Besides the spectral component at 3.25 [kHz], there are also components at 1.65, 4.9 and 6 [kHz].

The reason for additional frequency component occurrence can be found in the fact that there are more modes in the structure of the tool, which are excited by the pulse excitation used for testing [16].

It is imperative to balance responses of the amplifier and the sensor to see the influence of the complete measurement system. The total response is obtained by

Discontinuity Detection in the Vibration Signal of Turning Machines 37

Fig. 6 Recorded SNR of the amplifier used in the measurement system

serial connection of all the used sensors and amplifiers. The response is recorded or viewed at the PC screen or the oscilloscope.

Record of the signal to noise ratio (SNR) is performed to include the influence of the measurement system to the final results. Figure 6 is the result of the procedure. It is recorded in the system's idle. In the state of the idle there are no interactions between the tool and the insert. It can be concluded that the noise is − 80 [dB], which can be neglected, because it is not the significant impact.

Figure 7 shows the result of the system's testing to pulse extinction. The measured response approximately corresponds to the original manufacturer characteristics. The response frequency of the system is 21 [kHz].

5 Results

This section is divided into 2 parts:

- measured signals, and
- wavelet analysis of the measured signals.

The first subsection shows an example of the experimental results (the vibration signal). The second subsection deals with the analysis of the obtained vibration signals.

5.1 Measured Signals

Figure 8 shows the examples of the vibration signals from the experiments. It can be seen that signals are extremely non-stationary in its nature. Figure 8a shows the example of measured signals for the soft processing case. In the case

Fig. 7 Measurement's system response to the pulse extinction

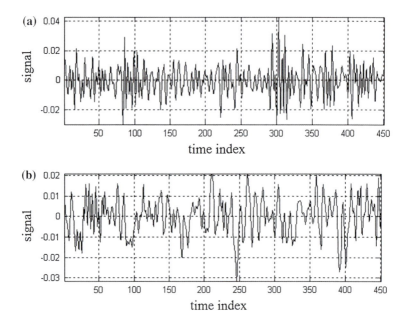

Fig. 8 Example of the vibration signals obtained in the experiments of: **a** soft processing, **b** hard processing

of hard processing, the example of the obtained results can be seen in Fig. 8b. It should be noted that the time index corresponds with real time if multiple by the sampling rate.

Discontinuity Detection in the Vibration Signal of Turning Machines

5.1.1 Soft Processing

The vibration signal in the soft processing is obtained at the tool's rotation speed of 1440 revolutions per minute.

The starting diameter of the insert was 40 [mm]. The diameter was 38.2 [mm] after the processing. Shift was 0.125 mm per revolution.

The average surface's roughness was 1.790 [μm] after the processing.

Figure 9 shows a time-dependent graph of the vibration signal obtained in the case of soft processing. 15 [s] is the time span for the signal acquisition.

There is no activity at the beginning of the signal for a 2 [s] time period. The actuators are engaged after 2 [s].

The contact between the insert and the tool is initiated at 3 [s] from the beginning of the signal. The insert and the tool are in the contact for additional 9.5 [s]. Then actuator is disengaged.

As can be seen, the signal's envelope is leveraged during the contact, which is expected in the case of the soft processing.

Figure 10 shows the spectrum of the corresponding signal vibration from Fig. 9. One can observe the activity in the signal's spectrum at the frequency range from 4 to 6 [kHz]. This activity can be explained as dynamics of cutting in the soft processing. The spectrum is approximately uniform at the frequencies between 2.5–12 [kHz].

5.1.2 Hard Processing

The signal vibration in the case of the hard processing is obtained at the rotation of 1120 revolutions per minute. The starting diameter of the insert was 40 [mm]. Diameter's insert was 38.2 [mm] after processing. The shift was 0.224 mm per revolution. The average surface's roughness was 1.996 [μm] after the processing.

Figure 11a shows the time-dependent representation of the vibration signal in the case of the hard processing. Figure 11b shows the corresponding spectrum.

Figure 11a shows that the contact between the tool and the insert is achieved at t = 2 [s]. The contact is terminated at t = 7.5 [s]. The signal's acquisition process has the duration of 10 s in total.

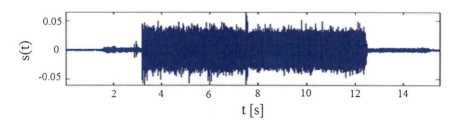

Fig. 9 Time domain representation of the vibration signal for the soft processing case

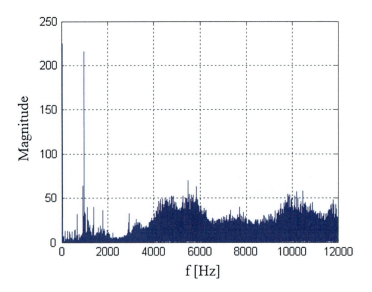

Fig. 10 Spectrum of the vibration signal in the soft processing

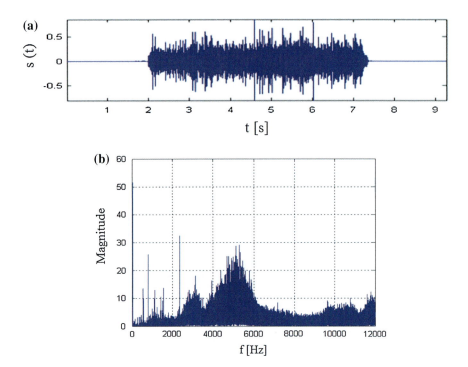

Fig. 11 The case of the hard processing: **a** time domain signal, **b** signal's spectrum

Uniformity of the graph points out that the signal's spectrum is also mostly uniform. Dynamics of cutting is exposed in the frequency interval from 4 to 6 [kHz]. The difference between the spectrum of the soft processing signal and the hard processing signal is in the magnitude of the frequency components.

5.1.3 Very Hard Processing

Experiments with very hard processing were performed at the rotation speed of the head of the tuning machine of 560 revolutions per minute. The start diameter of the insert was 43.5 [mm]. The diameter after the processing was 32.0 [mm]. The shift was 0.45 mm per revolution. The average surface roughness was 7.2 [μm]. Figure 12a shows the time-dependent representation of the vibration signal in this case. Figure 12b shows the corresponding spectrum.

The contact between the insert and the tool's head is established after 3 s from the start of the data acquisition process. Figure 12a shows that the envelope of the vibration signal changes considerable. There are two isolated parts of the signal, at the time locations $t = 1.7$ [s] and 2.5 [s]. The isolated areas exhibit the very large discrepancy in the signal's amplitude.

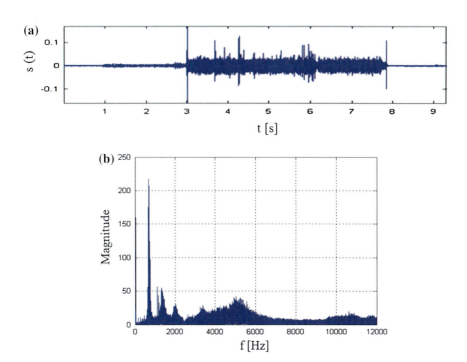

Fig. 12 The case of very hard processing: **a** time domain signal, **b** signal's spectrum

Figure 12b exhibits relatively high activity in the low-frequency part of the spectrum. There are two areas of high activity:

- from 700 to 2 [kHz] in the low-frequency band, and
- from 4 to 6 [kHz].

The second frequency range represents the natural frequency of the tuning tool.

At this point, it should be pointed out that the obtained vibration signals correspond in the quality to the referent signals from 'Bruel & Kjaer' or 'Kistler' sensors [1, 13].

5.2 Wavelet Analysis of the Measured Signals

The analysis of the vibration signal was performed by the discrete wavelet transform. The Daubechies wavelet of the 6th order was used for the wavelet decomposition.

Using the proposed algorithm explained in the third section, it is obtained that the mentioned wavelet is the best for the topic application. Figure 13a shows an example of the wavelet decomposition. The original signal is decomposed at 6 levels obtaining details and approximations at each level of the decomposition. Details coefficients are denoted as s1, s2, etc.

The approximation at the sixth level of the decomposition is denoted as a6.

Details are obtained by passing the signal through the low-pass filter. Approximations are obtained by passing the signal through high-pass filter. At the second level of the decomposition, instead of the original signal, approximation at the first level is used. The process is repeated several times. The number of levels depends on application, type of the wavelet and type of the signal and other conditions. After all, it can be chosen arbitrary, depending on heuristic knowledge or what can be lost in the processing. Finally, sets of the wavelet coefficients are obtained.

Figure 13b shows the scaling and wavelet functions of the db6 wavelet. Furthermore, discrete decomposition low-pass and high-pass filters are shown. Reconstruction filters are shown at the bottom of the Fig. 13b.

The fundamental properties of the db6 wavelet are:

- orthogonality, and
- five moments are equal to zero, which means that it approximates the polynomial of the fifth order very well in the approximation coefficients at the last level of decomposition.

Figure 14 shows an example of the obtained results by illustrating how wavelets detect a discontinuity in the soft processing of inserts by turning machine.

Figure 14a presents signal, approximation coefficients at level 5 and 5 levels of wavelet detail coefficients. Therefore, it can be concluded that the output set of the

Discontinuity Detection in the Vibration Signal of Turning Machines 43

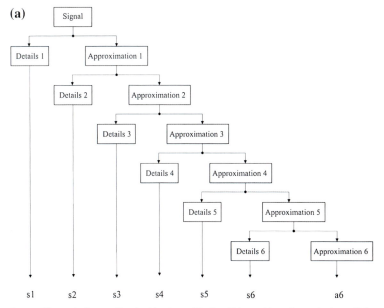

Outputs of the wavelet decomposition algorithm: 6 sets of detail coefficients and 1 set of approximation coefficients

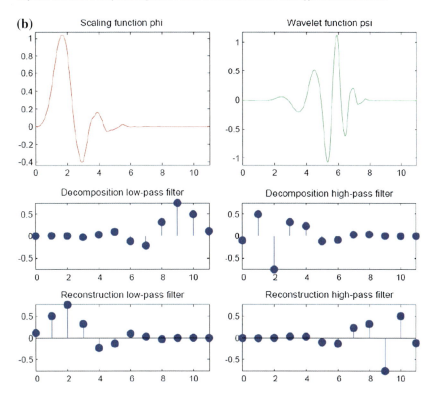

Fig. 13 Wavelet decomposition: **a** output sets of the wavelet decomposition algorithm, **b** scaling and wavelet functions for the db6 wavelet

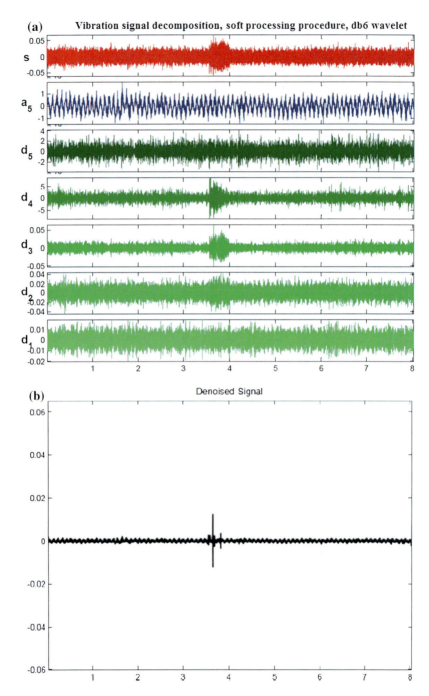

Fig. 14 Example of the results (soft processing): **a** wavelet decomposition, **b** detection in denoised signal

coefficients consists of the approximation coefficients at level 5 and 5 levels of wavelet detail coefficients.

The problem with the representation in Fig. 14a is in the fact that only experienced eyes can discern detection in this manner of presentation. Therefore, the signal is denoised by the DWT and the thresholding method. The result is very clear and shown in Fig. 14b for one decomposition set.

It should be notated that the y-axis values are the magnitudes of the wavelet coefficients, which are relative values of the correlation in the wavelet domain. However, X-axis has physical units: [kHz].

The vibration signal (soft processing, Fig. 14a) is decomposed at the five levels due to dynamics of cutting. Namely, the dynamics of cutting is dominant in the frequency range between 3 and 6 [kHz]. The mentioned frequency range corresponds to the details coefficients at the third and forth levels (d4 and d3). One can see the change of amplitude in the details coefficients in the frequency range between 3.5 and 4 [kHz].

Due to uniformity of the spectral response, it is not possible to make conclusions about existence of the discontinuity in the mentioned frequency range.

The method of the hard thresholding is applied to the details coefficients d3 and d4. The threshold is set to 85 % of the peak value for the corresponding coefficients at every level of the decomposition. All remaining coefficients at all the levels of the decomposition are discarded. The final result is equal to the approximation and details d3 and d4. The result of such processing is shown in Fig. 14b.

After the hard thresholding, it can be observed that a discontinuity occurs at the position 3.5. It can be concluded that the vibration signal obtained by the soft processing must be thresholded by the hard threshold if we wish to detect discontinuities. The reason is in the high rotation speed of the tuning machine and small shift. If the insert and the tool separate, the separation does not manifest through the entire spectrum, but only through a part of it.

The research is performed only in the frequency range between 3 and 6 [kHz], because the main dynamics of the cutting is in that range. If some deformation occurs on the insert's surface during the processing, the anomaly is not visible in the entire spectrum.

Figure 15 shows an example of the results of wavelet detection for the hard processing case.

The vibration signal has a slightly different frequency range of the cutting dynamic: it is between 2.5 and 6 [kHz]. However, the intensity of the signal is increased in the entire spectrum. It can be seen from the graphical representation of the detail coefficients that there are considerable contributions to the details d1 and d2 at the positions 1.2 and 1.4. It can be concluded that the discontinuities exist at these positions. If we observe the approximation a5, we can see a considerable increase in the coefficients' value at the mentioned positions. The reason is in the hard processing, because of the velocity of the tuning machine's head, which is smaller than in the soft processing.

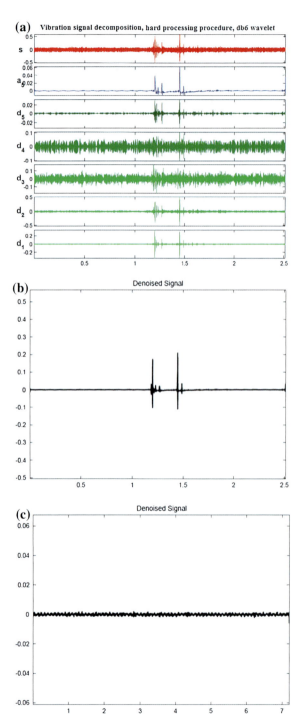

Fig. 15 Example of the results (hard processing): **a** wavelet decomposition, **b** detection in the denoised signal (soft thresholding), **c** detection in the denoised signal (hard thresholding)

Discontinuity Detection in the Vibration Signal of Turning Machines

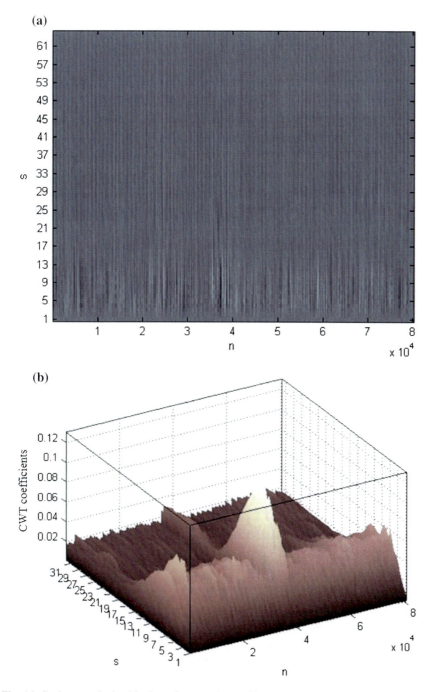

Fig. 16 Scalogram obtained in the soft processing: a 2D representation, b 3D representation

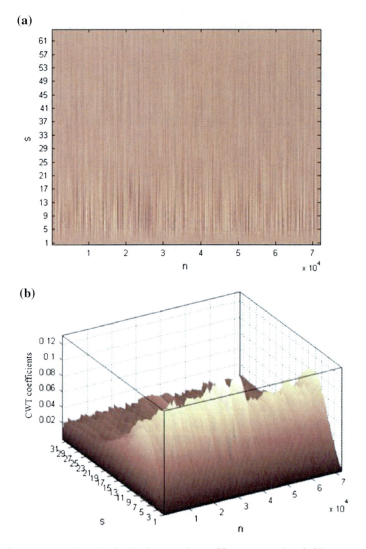

Fig. 17 Scalogram obtained in the hard processing: **a** 2D representation, **b** 3D representation

However, more materials are removed from the insert in the hard processing. If the tool and the insert separate, the damage of the surface manifests itself by the occurrence of the discontinuities. These discontinuities are precisely located by the DWT after the soft thresholding procedure. Figure 15c shows what happens if we use hard thresholding in the case of hard processing (there are no detections of the discontinuities).

From the Fig. 15b, it can be seen that there are two dominant discontinuities and smaller discontinuities in the neighborhoods. Finally, it can be concluded that the tool and the insert separate in the hard processing of the used metal.

Discontinuity Detection in the Vibration Signal of Turning Machines

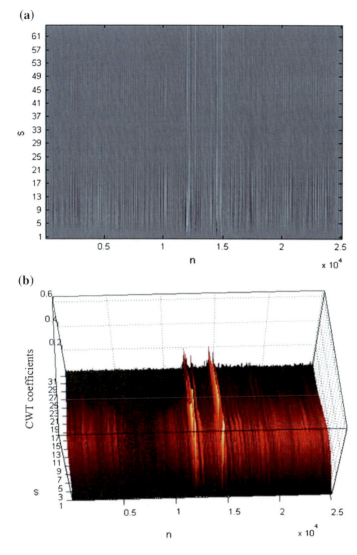

Fig. 18 Scalogram obtained in the very hard processing: **a** 2D representation, **b** 3D representation

In both cases (Figs. 14 and 15) it can be seen that denoising of the signal plays a vital role in the understanding of the results.

Figure 16 shows the scalogram of the soft processing around t = 8.7 [s]. Figure 16a shows a 2D representation and Fig. 16b a 3D representation.

Figure 17 shows the scalogram for the hard processing.

Figure 18 shows the same for the very hard processing.

Figure 19 shows the scalogram for the very hard processing with a different vibration signal.

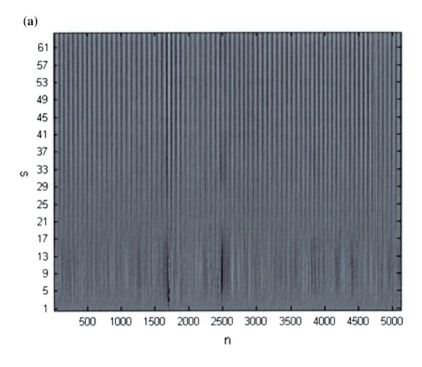

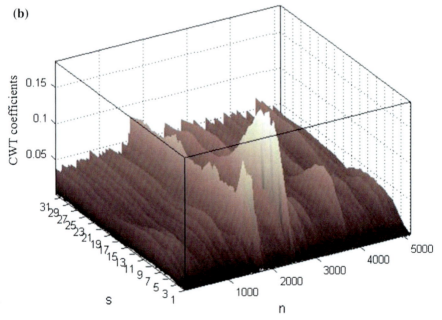

Fig. 19 Scalogram obtained in the very hard processing (different signal): **a** 2D representation, **b** 3D representation

6 Conclusion

Discontinuities in the vibration signal are detected by the application of the DWT. Research has shown that the wavelet selection influences DWT results. Wavelets greatly influence the classification of results. Correct wavelet selection improves diagnostics. It also greatly influences real-time performance, due to the rapid convergence of the DWT.

The location of the discontinuity in the vibration signal correlates with the occurrence of damage on the surface of the processed insert. In the chapter, the possibility of precise localization of the discontinuity in the vibration signal is confirmed in case of wavelet usage. Therefore, better diagnostics are achieved. Significant economic savings are made possible by the procedure.

The chapter can be summarized into four points:

- DWT selection methodology (due to this being a vibration signal and quality control is performed by the means of surface roughness);
- wavelet selection methodology (spectrum minimization);
- signal filtering in search of holes influencing quality control;
- approximation signal represents the appearance of the metal surface after turning, provided that the signal is obtained from the sufficient number of samples.

A.1 7 Appendix 1: General Mathematical Conditions for the Selection of the Wavelet

There are four conditions, which have to be fulfilled by the scaling function, $\varphi(t)$, in order to be considered as the solution:

1. for $t \in R$: $\|\phi(t)\| = 1$
2. $\sum_{k} \left|\hat{\phi}(\omega + 2k\pi)\right|^2 = 1$
3. $\lim_{j \to \infty} \hat{\phi}(2^{-j}\omega) = 1$
4. $\hat{\phi}(2\omega) = \hat{H}(\omega) \cdot \hat{\phi}(\omega)$

where:

- $\hat{\phi}$ is the Fourier transform of the function φ,
- $\hat{H}$ is the frequency response of the low-pass filter in form:

$$\hat{H}(\omega) = \sum_k h(k) \cdot e^{-i\omega k}. \qquad (13)$$

Furthermore, $\varphi(t)$ is defined for the orthonormal case in $L_2(R)$ vector space.

The first condition is the general condition of the orthogonality. The second condition is the Poisson's equation, which is equal to orthogonality property for the shifted scaling function, $\varphi(t-k)$, where $k \in Z$. The third condition is the un-interruption property (the function is continuous) in $\omega = 0$. The fourth condition is the so called dilatation equation.

It should be notice that the low-pass filter, H, and the high-pass filter, G, should be in the following relationship for all frequencies:

$$\hat{G}(\omega) = e^{i\omega} \cdot \bar{\hat{H}}(\omega + \pi). \qquad (14)$$

Functions $\hat{G}$ and $\hat{H}$ are the periodic functions with period 2π.

A.2 8   Appendix 2: Conditions for the Amplitude and Phase of the Chosen Wavelet's Spectrum

By the multiresolution theory, the spectrum of the frequency limited scaling function has the compact domain in the interval $\langle -\omega_m, \omega_m \rangle$, where $\omega_m = \pi + \alpha$ and $0 \leq \alpha \leq \frac{\pi}{3}$.

The corresponding wavelet function, $\left|\hat{\psi}(\omega)\right|$, can be expressed with the scaling function with the substitution:

$$g(\omega) = \left|\hat{\phi}(\omega)\right|. \qquad (15)$$

The obtained expression is more appropriate for further considerations:

$$\left|\hat{\psi}(\omega)\right| = \begin{cases} 0 & 0 \leq |\omega| < \pi - \alpha \\ g(2\pi - \omega) & \pi - \alpha \leq |\omega| < \pi + \alpha \\ 1 & \pi + \alpha \leq |\omega| < 2\pi - 2\alpha \\ g\left(\frac{\omega}{2}\right) & 2\pi - 2\alpha \leq |\omega| < 2\pi + 2\alpha \\ 0 & 2\pi + 2\alpha \leq |\omega| \end{cases} \qquad (16)$$

From the above equation, it can be concluded that $g(\omega)$ is a positive function. The orthonormal wavelet defined with Eq. (16) has the compact domain, which varies due to parameter α selection. For $\alpha = 0$, frequency is the interval between π and 2π. For $\alpha = \frac{\pi}{3}$, the frequency range is: $\left[\frac{2\pi}{3}, \frac{8\pi}{3}\right]$.

The necessary condition for the chosen wavelet is the value of the parameter α, which must be equal to $\frac{\pi}{3}$ to satisfy the condition for the amplitude.

In order to derive the phase condition, we start with the relationship between the wavelet and the scaling function:

$$\hat{\psi}(2\omega) = e^{i\omega} \cdot \bar{\hat{H}}(\omega + \pi) \cdot \hat{\phi}(\omega) \tag{17}$$

Equation (17) can be written as Eqs. (18–20):

$$\hat{H}(\omega) = \frac{\hat{\phi}(2\omega)}{\hat{\phi}(\omega)} \tag{18}$$

$$\hat{H}(\omega + \pi) = \frac{\hat{\phi}(2\omega + 2\pi)}{\hat{\phi}(\omega + \pi)} \tag{19}$$

$$\bar{\hat{H}}(\omega + \pi) = \frac{\bar{\hat{\phi}}(2\omega + 2\pi)}{\bar{\hat{\phi}}(\omega + \pi)} \tag{20}$$

Combining Eqs. (17) and (20), it can be written:

$$\hat{\psi}(2\omega) = e^{i\frac{\omega}{2}} \cdot \frac{\bar{\hat{\phi}}(2\omega + 2\pi)}{\bar{\hat{\phi}}(\omega + \pi)} \cdot \hat{\phi}(\omega) \tag{21}$$

or:

$$\hat{\psi}(\omega) = e^{i\frac{\omega}{2}} \cdot \frac{\bar{\hat{\phi}}(\omega + 2\pi)}{\bar{\hat{\phi}}\left(\frac{\omega}{2} + \pi\right)} \cdot \hat{\phi}\left(\frac{\omega}{2}\right) \tag{22}$$

From Eq. (22), the phase of the wavelet function can be written as:

$$\theta_\psi(\omega) = \frac{\omega}{2} - \theta_\phi(\omega + 2\pi) + \theta_\phi\left(\frac{\omega}{2} + \pi\right) + \theta_\phi\left(\frac{\omega}{2}\right) \tag{23}$$

where:

- θ_ψ is the phase of the $\hat{\psi}$ function, and
- θ_ϕ is the phase of the $\hat{\phi}$ function.

Due to periodicity of the $\hat{H}$ function, the phase of the $\hat{\phi}$ function satisfy the following equation:

$$\theta_\phi(2\omega) - \theta_\phi(\omega) + \theta_\phi(4\pi - 2\omega) - \theta_\phi(2\pi - \omega) = 0 \tag{24}$$

Equation (24) is the phase condition for the wavelet choice.

References

1. Dimla Sr, D.E.: The impact of cutting conditions on cutting forces and vibration signals in turning with plane face geometry inserts. J. Mater. Process. Technol. **155–156**, 1708–1715 (2004)

2. Scheffer, C., Heyens, P.S.: Wear monitoring in turning operations using vibration and strain measurements. Mech. Syst. Signal Process. **15**, 1185–1202 (2001)
3. Grzesik, W., Brol, S.: Wavelet and fractal approach to surface roughness characterization after finish turning of different workpiece materials. J. Mater. Process. Technol. **209**, 2522–2531 (2008)
4. Šoda, J.: Wavelet transform based discontinuities detection of vibration signal. PhD Thesis, University of Split, Faculty of Electrical Engineering, Mechanical Engineering and Naval Architecture (2010)
5. Dimla Sr, D.E.: Sensor signals for tool-wear monitoring in metal cutting operations—a review of methods. Int. J. Mach. Tool Manuf. **40**, 1073–1098 (2000)
6. Jansen, M., Oonincx, P.: Second Generation Wavelets and Applications. Springer, London (2005)
7. Vujović, I., Šoda, J., Kuzmanić, I.: Cutting-edge mathematical tools in processing and analysis of signals in marine and navy. Trans. Marit. Sci. **1**, 35–48 (2012)
8. Mallat, S.: A Wavelet Tour of Signal Processing. Academic Press, New York (2009)
9. Selesnick, I.W., Baraniuk, R.G., Kingsbury, N.G.: The dual-tree complex wavelet transform. IEEE Signal Process. Mag. **22**, 123–151 (2005)
10. Christopher, H., Walnut, D.F.: Fundamental Papers in Wavelet Theory. Princeton University Press, London (2006)
11. Kuzmanić, I., Šoda, J., Antonić, R., Vujović, I., Beroš, S.: Monitoring of oil leakage from a ship propulsion system using IR camera and wavelet analysis for prevention of health and ecology risks and engine faults. Mat.-wiss. u Werkstofftech **40**, 178–186 (2009)
12. Polikar, R.: Lecturers notes. Rowan University. http://www.engineering.rowan.edu/~polikar/CLASSESS/ECE554 (2002)
13. Risbood, K.A., Sahasrabudhe, U.S.: Prediction of surface roughness and dimensional deviation by measuring cutting forces and vibrations in turning process. J. Mater. Process. Technol. **132**, 203–214 (2003)
14. Sun, J., Rahman, M., Wong, Y.S., Hong, G.S.: Multiclassification of tool wear with support vector machine by manufacturing loss consideration. Int. J. Mach. Tool Manuf. **44**, 1179–1187 (2004)
15. Alvin, K.F., Robertson, A.N., Reich, G.W., Park, K.C.: Structural system identification: from reality to models. Comput. Struct. **81**, 1149–1176 (2003)
16. Benardos, P.G., Vosniakos, G.: Predicting surface roughness in machining: a review. Int. J. Mach. Tool Manuf. **43**, 833–844 (2003)

Visualization of Global Illumination Variations in Motion Segmentation

Igor Vujović, Ivica Kuzmanić, Joško Šoda
and Slobodan Marko Beroš

Abstract Motion segmentation is the most important part in many applications, such as surveillance, security, monitoring, recognition, etc. The presented research deals with short-term illumination variations in video streams. Illumination variations influence values of pixels and greatly impact the segmentation mask obtained as a part of a motion detection algorithm. In order to subjectively visualize the extent of variations, these must be emphasized. The chapter presents a wavelet energy model based algorithm which detects and emphasizes illumination variations.

Keywords Wavelet transform · Motion detection · Illumination variation

1 Introduction

Global illumination changes are a non-stationary phenomenon preventing correct segmentation of motion. This presents a problem in any application using video stream, i.e. surveillance, quality control, monitoring of engines or processes, vision-based control, pattern or face recognition, etc.

I. Vujović · I. Kuzmanić (✉) · J. Šoda
Faculty of Maritime Studies, University of Split, Zrinsko-Frankopanska 38,
21000 Split, Croatia
e-mail: ikuzman@pfst.hr

I. Vujović
e-mail: ivujovic@pfst.hr

J. Šoda
e-mail: jsoda@pfst.hr

S. M. Beroš
Faculty of Electrical Engineering, Mechanical Engineering and Naval Architecture,
University of Split, Ruđera Boškovića 32, 21000 Split, Croatia
e-mail: sberos@fesb.hr

The phenomenon may be caused by a number of things such as:

- camera's self adjust or other technical reasons,
- normal oscillations in source intensity,
- object that covers a part of the image or source affecting global and/or local illumination change, etc.

For example, oscillations of a natural source may be caused by a "light production process" in the sun, atmospheric conditions or occurrences. Variations in illumination of an artificial source may be caused, for example, by variations of network voltage.

The problem of illumination variations is real in both outdoor and indoor applications [1, 2]. Illumination variations present a problem even in security applications, as shown for example in [1], where face recognition is considered. Robust character region extraction is under influence of variations in illumination, as shown in [2]. Suppression of illumination variations is discussed in many papers, e.g. in [3, 4] due to their influence on algorithm performance. However, visualizing the variations by brushing [5] or some other technique is unusual.

The problem with defining variations is that they occur in both the foreground and the background. Therefore, selecting proper threshold is not enough for visualization. It would be ideal to choose different thresholds for background and foreground, but due to illumination variations the borders between them are impossible to establish by automated methods, leaving us with a vicious circle. On the other hand, manual segmentation is not acceptable due to speed and extent work required for a human operator.

Change detection algorithms have been proposed in order to reduce the effect of varying illuminations [3, 6–9]. Spatially invariant models are produced by applying change detection algorithms on analyzed visual data. These models reduce illumination variations and consequently reduce the effects of sudden illumination changes. The algorithm's drawback is that it produces double detections. Another approach, a hybrid strategy [3] has been proposed and implemented in order to reduce illumination variations. The advantage of using a hybrid strategy is efficiency of the algorithm. The main drawback is inefficient background removal.

In order to suppress illumination variations, an adaptive background mixture model is presented in [6]. The application of the proposed algorithm was in real-time tracking.

Anomalous in motion detection caused by illumination variations are considered in [7]. Change detectors of two types were quantitatively compared. The change detectors were based on a quadratic covariance.

The influence of the variations in illumination to the motion detection is covered in [8]. The change detection approach was applied. However, this was applied to a low frame-rate video due to application constrains in computer power.

The influence of change detection to the global thresholding in the image (frame) is considered in [9].

However, these models are not used to isolate illumination variations. It can be said that these models detect variations in illumination in implicit form. Moreover, these models do not extract illumination variations and annul their effect. Rather it can be said that these models use statistics to annul the data for which there is a reasonable doubt that there are illumination variations.

The simple reason for visualization of variations in illumination is to see how they look like, and eventually, subtract them from the foreground. The first stage is to see their influence on the foreground, which is the scope of the chapter. Improvements are made by wavelet transform (WT) or discrete wavelet transform (DWT) to be precise.

The DWT is used for purposes of achievement of computer implementation efficiency. The DWT is implemented by lifting steps to further increase execution speed and this implementation is called lifting wavelet transform (LWT). The lifting implementation is significantly faster than filter-banks implementation of WT due to simplified calculations. The simplifications allowed the development of programming solutions capable of being executed much faster on a digital computer.

The chapter is organized as follows.

In the second section we discuss some implementation issues of the theory of wavelets and a model of illumination variations.

The third section describes the proposed technique and algorithm steps.

Results are presented in the fourth section along with experimental equipment and mathematical expressions necessary to understand and interpret the results.

The final section is the conclusion.

2 Theory

In this section, two issues are dealt with:

- wavelet transform as a tool for the execution of task, and
- the mathematical model of variations.

Like in any research, it is important to choose tools appropriately and to explain others why to use the chosen tool. Therefore, the first part of the Sect. 2.1 explains the choice of transform and theory regarding it.

To achieve the results, a model of variations should be introduced. The model should use mathematics and should be programmable and implementable in the digital computer. Therefore, the second part of the Sect. 2.2 explains the model used.

2.1 Wavelets

The basic theory of wavelets is well known and presented in numerous publications, e.g. in [10–12] and there is no need for repetition. Since some new transforms have been introduced in the last decade, it is important to establish the reason for not using some other transform for the particular problem. The comparison of new wavelet-like transforms and wavelets, with application in video surveillance is presented, explained and illustrated in [13]. The reason is explained in more details at the end of the Sect. 2.1.

When a problem appears, the first step is to choose a manner of its solution. For example, one might ask oneself why to use wavelets. Wavelets exhibit huge success in one-dimensional problems of signal processing and analysis. However, the problem of illumination variation can be considered one-dimensional if only the dimension of time is taken into account. In that case, every pixel is independent of its neighbors and its value is time-dependant. This approach can be promising if we are merely interested in image processing, instead of a deep analysis. This is due to the fact that some complex scene situations can result in space-dependence of pixel values. Therefore, multi-dimensional dependence occurs.

Therefore, two types of LWT or DWT can be used:

- two-dimensional (2D) or
- two-dimensional plus time (2D + T) as the third dimension.

It should be noted that 2D LWT or 2D DWT are used in image processing applications. 1D LWT or DWT is used for data which have time and, for example, amplitude. That means 1D is actual a two-dimensional and 2D a three-dimensional transform.

When using 2D LWT, indexed time should be incorporated by some other way, such as in memory-based motion detection algorithms [4, 14, 15] or some sort of Kalman filtering. In memory-based motion detection algorithms accumulator of frames is used to average the influence of illumination variations [15]. Such algorithms have the advantage of faster execution after initialization. Four wavelet techniques for the improvements are introduced in [14]. The advantage of the 2D approach is in fact that there are several standard open source, standard or commercial codes available in evaluation of such algorithms [16].

Figure 1 illustrates a MBMD method. It can be seen that time is incorporated by averaging operation.

On the other hand, 2D + T methods are more suitable for off-line applications. In such a case, it is often the case that recorded video sequence is frequently analyzed in its entirety. 2D + T methods have less publicly available open source codes, making it harder for researchers to compare and evaluate new and older algorithms [17]. 2D + T transforms can be realized in two manners:

- time decomposition first, and
- spatial decomposition first.

Figure 2 illustrates the "time first" method.

In the "time first" procedure, decomposition is executed in the time dimension and then in spatial dimension. Spatial decomposition results in approximation and

Visualization of Global Illumination Variations

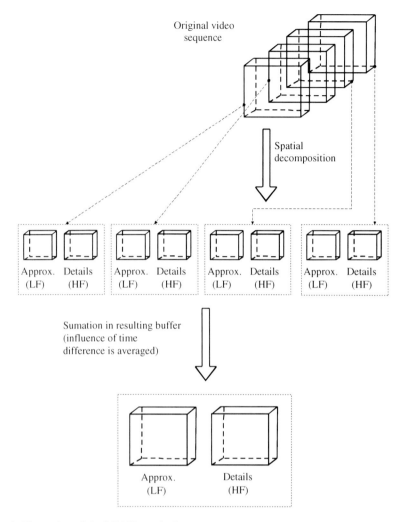

Fig. 1 Illustration of the MBMD method

details. In case of the wavelet transform, a spatial decomposition results in the corresponding low-frequency and high-frequency coefficients.

Advances in science and technology have brought about new transforms suitable for some purposes, which could be used in this application as well. Generally speaking, there are several new transforms which could be used for image processing, which have proved to be better than wavelets in the considered applications [18–21], but are not better in general. There are wavelet-inspired transforms with wrapped or rotated wavelet basis. If different tilling is used, other transforms are obtained. For example [13, 22]:

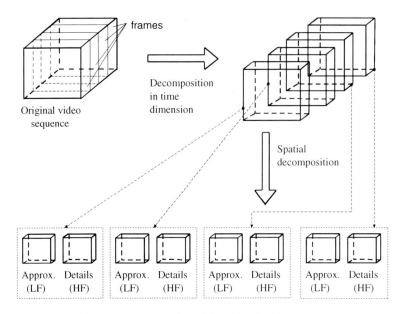

Fig. 2 Illustration of 2D + T decomposition of the video signal

- a directional wavelet transform is obtained if we divide the corona into a constant number regardless of scale,
- a ridgelet transform is obtained by subdividing each dyadic corona into $C \cdot 2^j$ angles, where 2^j is the scale of the transform and C the corresponding coefficient,
- a Gabor analysis is obtained if we substitute coronae with fixed width for dyadic coronae.

Furthermore, an adaptive partitioning of the frequency plane can be created, which best matches the features of the analyzed signal. This is the construction of so called ridgelet packets [13].

However, this could be seen as using a sledge hammer to kill an ant.

Yet another problem is the issue of the duration of a desired operation. If the transform requires significantly more time to process than the WT, then the WT could be considered nevertheless. This is due to need to operate in real-time conditions and with the considerable frame rate.

However, the duration problems of new wavelet-inspired transforms are not caused by extensive operations required, but rather by programming solutions. It could be expected that these issues will be solved in future.

2.2 Illumination Variations Model

Several issues have to be taken into account when developing the model of illumination variations. The first one seems trivial at first glance.

The first issue is how to define the phrase short-term in the variation illumination phenomenon. It can be defined in terms of time or number of frames. The latter depends on the computer's clock, which differs from computer to computer. The first, time, is appropriate for physical evaluation and is in absolute measure. However, computers operate in discrete time, which is the number of frames (or indexed time). If we consider an average present-day PC, illumination variation can last one or two frames.

Of course, another variation can occur almost immediately. So, there is no certain ground truth. This is a problem, because many scientists ask for comparison with publicly available ground truth data. However, they are unaware that all video-sequences are contaminated with variation illumination.

The human visual system can compensate for variation illuminations and we see a "clear" image. However, machine/robot/computer vision algorithms cannot compensate for illumination variations, which greatly influence their performance.

Illuminations are modeled in [4, 23, 24].

The image gray value level $I(x, y)$ is assumed to be proportional to the product of reflectance $r(x, y)$ and illumination $e(x, y)$ [24]:

$$I(x, y) = r(x, y) \times e(x, y) \qquad (1)$$

Of course, Eq. (1) is a way to describe any arbitrary or random noise. There is no way to calculate the actual noise except experimentally, by establishing the differences between supposedly identical images.

In case of color images, the matrix I has an additional dimension and Eq. (1) can be modified as follows:

$$I(x, y, c) = r(x, y, c) \times e(x, y, c) \qquad (2)$$

where c is the color dimension. In the case of a RGB color space, every plane of I is influenced by variation illuminations. In the case of the HSI color space, only the intensity plane is influenced by variations. Therefore, the number of operations will be reduced to 1/3 if HSI is used instead of RGB.

Therefore, the choice of the color-space could be an issue.

For further consideration, it is assumed that one of the two neighboring frames is free of variations in illumination. Additive Gaussian noise is present in both frames. In this case, the two images can be expressed by the following equation introduced in [1]:

$$I(x, y, n) = S(x, y, n) + \eta(x, y, n) \qquad (3)$$

$$I(x, y, n+1) = S(x, y, n+1) + \eta(x, y, n+1) + \zeta(x, y, n+1) \qquad (4)$$

where:

- n is the time index,
- $I(n)$ is an nth frame (frame in time t),
- $I(n + 1)$ is an $(n + 1)$th frame (frame in time $t + \Delta T$),
- $S(n)$ is the projection of the scene onto the image plane in the nth frame (nth time index)
- $S(n + 1)$ is the projection of the scene onto the image plane in the $(n + 1)$th frame (or corresponding time index),
- $\eta(n)$ is the additive Gaussian noise in the nth frame,
- $\eta(n + 1)$ is the additive Gaussian noise in the $(n + 1)$th frame, and
- $\xi(n + 1)$ is an additive and a multiplicative illumination variation.

The complexity of situation can be seen if Eqs. (3) and (4) are rewritten for color spaces:

$$I(x, y, c, n) = S(x, y, c, n) + \eta(x, y, c, n) \tag{5}$$

$$I(x, y, c, n+1) = S(x, y, c, n+1) + \eta(x, y, c, n+1) + \zeta(x, y, c, n+1) \tag{6}$$

It can be seen that I depends on four variables:

- 2 space coordinates,
- one color dimension, and
- time frame.

Additive and multiplicative variations consist of two components [1]:

$$\zeta(x, y, c, n+1) = a + b \times S(x, y, c, n+1) \tag{7}$$

where:

- a can be considered as DC-offset or constant brightness change (positive or negative) at global level,
- the constant b describes the extent of global changes in illumination.

Therefore, image differences should exhibit the phenomenon of artifacts. The $(n + i)$th frame can be described as:

$$I(x, y, c, n+i) = S(x, y, c, n+i) + \eta(x, y, c, n+i) + \zeta(x, y, c, n+i) \tag{8}$$

If all the images are supposed to be the same (as in the case of a stationary scene), then:

$$S(x, y, c, n) = S(x, y, c, n+1) = S(x, y, c, n+i) \tag{9}$$

By summing all the images into the so called training sequence, we obtain:

$$\sum_{i=0}^{N-1} I(x,y,c,n+i) = \sum_{i=0}^{N-1} S(x,y,c,n+i)$$
$$+ \sum_{i=0}^{N-1} \eta(x,y,c,n+i) + \sum_{i=0}^{N-1} \xi(x,y,c,n+i) \qquad (10)$$

and since:

$$\sum_{i=0}^{N-1} S(x,y,c,n+i) = N \cdot S(x,y,c,n) \qquad (11)$$

It can be written:

$$\sum_{i=0}^{N-1} I(x,y,c,n+i) = N \cdot S(x,y,c,n)$$
$$+ \sum_{i=0}^{N-1} \eta(x,y,c,n+i) + \sum_{i=0}^{N-1} \xi(x,y,c,n+i) \qquad (12)$$

For capital N (large N means large training sequence for stationary scene), illumination variations are averaged to zero:

$$\sum_{i=0}^{N-1} \xi(x,y,c,n+i) = 0 \qquad (13)$$

Therefore, illumination variations are eliminated and Eq. (12) reduces to the following form:

$$\sum_{i=0}^{N-1} I(x,y,c,n+i) = N \cdot S(x,y,c,n) + \sum_{i=0}^{N-1} \eta(x,y,c,n+i) \qquad (14)$$

An analogue analysis can be applied to the expression for noise, $\sum_{i=0}^{N-1} \eta(x,y,c,n+i)$, which has zero-mean variance [25].

As can be concluded, the phenomenon of interest disappears if a large sequence is used in a stationary scene.

The problem cannot disappear in a dynamic scene or when using a moving camera, such as in satellites, robots, autonomous vehicles, etc. This fact makes it even more difficult to find a solution if a dynamic scene of moving camera is used. Therefore, the proposed technique deals only with stationary camera problems.

In order to use a moving camera, a sort of camera compensation could be integrated in the proposed technique as additional module. Alternatively, large and small accumulators can be used:

- the large one instead of the background model and
- the small one instead of the accumulator for illumination variations suppression.

This could be a way for further researches.

3 Proposed Technique

Parseval relation, which links energy in different domains, is the basis of the proposed technique. In short, Parseval proved that time convolution (integration in time domain) can be replaced by the summation of squared frequency components of the spectra (frequency domain) or by the inner product (in wavelet domain) and the result will be the same. Particularly, the relation deals with the calculation of power/energy.

Since there are several definitions of this relation, depending on application area and domain used, we will use Definition 3.1 that covers all important variations of the relation.

Standard signal processing nomenclature is used in the definition.

Definition 3.1 Reference [4] If f and h are in $L^1(R) \cap L^2(R)$, then following definitions are valid:

1. Parseval's expression is:

$$\int_{-\infty}^{+\infty} f(t)h^*(t)dt = \frac{1}{2\pi} \int_{-\infty}^{+\infty} \hat{f}(\omega)\hat{h}^*(\omega)d\omega \tag{15}$$

where $\hat{f}$ and $\hat{h}$ are Fourier's transformants of $f(t)$ and $h(t)$.

2. If $f = h$ follows so called Plancher's expression:

$$\int_{-\infty}^{+\infty} |f(t)|^2 dt = \frac{1}{2\pi} \int_{-\infty}^{+\infty} |\hat{f}(\omega)|^2 d\omega \tag{16}$$

3. Energy of signal can be calculated from the time-domain as well as from the frequency-domain as an integral in time interval or as sum of energies of discrete spectrum components, if discrete Fourier transform (DFT) is known:

$$\frac{1}{2\pi} \int_{-\pi}^{\pi} \left[\sum_{n=-\infty}^{+\infty} f(n) e^{-j\omega n} \right] F^*(e^{j\omega n}) d\omega = \sum_{n=-\infty}^{+\infty} |f(n)|^2 \tag{17}$$

4. If the properties of the frequency convolution of the DFT are used, we arrive at the following expression:

$$\sum_{n=0}^{N-1} f(n)f(n)e^{-jkn} = \frac{1}{N} \sum_{i=0}^{N-1} F(i)F(k-i) \tag{18}$$

If we assume that $k = 0$, it can be stated that:

$$\sum_{n=0}^{N-1} f^2(n) = \frac{1}{N} \sum_{i=0}^{N-1} F(i)F(-i) = \frac{1}{N} \sum_{i=0}^{N-1} |F(i)|^2 \tag{19}$$

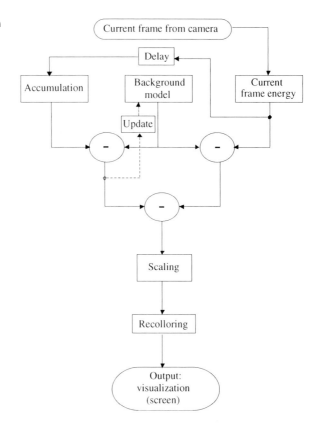

Fig. 3 Overall flow diagram of the proposed technique

These results allow the density spectra of discrete energy or the estimate of the spectral pediogram to be defined as:

$$S(k) = |F(k)|^2 \quad \text{for } 0 \leq k \leq N - 1 \tag{20}$$

The idea of illumination variation visualization originates from the motion detection algorithm with the suppression of the previously mentioned variations [4]. If the algorithm from [4] can suppress the variations, the logical conclusion is that variations can be detected in an indirect manner. Therefore, the process of variation detection should be used, but the opposite should be done after the detection. Instead of suppression, the variations should be emphasized.

Figure 3 shows the overall flow diagram of the proposed technique.

The input of the proposed algorithm is the camera data obtained in a present time frame denoted with n.

The initialization of the algorithm includes an analysis of the scene for the formation of the background model and the initial accumulator. The background model is formed by calculating the average energy for every position in the scene without motions. It is explained in Sect. 2.2 that averaging eliminates noises and illumination variations. Therefore, the background model is a variation-free and noise-free representation of a static scene when there are no motions.

It is important to have a variation-free representation of the scene to detect the variations in illumination.

Using Parseval relation, the single pixel energy of all frames at position k in training sequence is calculated as [4]:

$$E(x,y,c) = \frac{1}{N}\sum_{k=1}^{N}\left(|I_{w1}(x,y,c,k)|^2 + |I_{w2}(x,y,c,k)|^2\right) \quad (21)$$

where x, y are spatial coordinates of pixel, c designation of the color, and k is a discrete-time index which is equal to the frame sequence number in training sequence. I_{w1} is an approximation coefficient obtained from applying the DWT transform with mother wavelet 1 at position (x, y) of frame k, and I_{w2} is an approximation coefficient obtained from applying the DWT transform with mother wavelet 2 at position (x, y) of frame k.

Since only general indication "mother wavelet 1" and "mother wavelet 2" are used, it should be noted that the problem of choosing wavelets is a topic which could be elaborated both in an intuitive and mathematical manner. Both manners require significant space and could raise different issues and controversies. This is a sensitive matter and includes the area of application and specific conditions which the application requires. It could be defined by a number of vanishing moments in a chosen mother wavelet. Therefore some space should be devoted to discussing the choice.

Two wavelets can be chosen to meet different needs, such as in [26], where the percentage of the correct detections (both foreground and background pixels) in the total number of pixels in the frame and time of execution were the two criteria considered.

By analogy, one criterion could be illumination variation detection and the other duration. The optimum wavelet selection solution could be arrived at by ranking mother wavelet pairs according to both criteria and determining the total best rank [24]. Another approach is to use an error minimization procedure, such as the well-known mean square error (MSE) or its equivalent.

The referent single pixel energy at position (x, y) is determined by normalizing (21) with the maximum value of E. Matrix E is the 3-D. It consists of pixel energies. It is built in three layers expressed by Eqs. (22, 23) and (24).

Equation (22) represents the red channel energies by pixels, where the value of the position (1, 1) represents the red channel energy for the pixel at the position (1, 1) in the corresponding image. The value in the position (1, 2) represents the red channel energy for the pixel at the position (1, 2) in the corresponding image, etc.

Therefore, Eq. (22) may be expressed as matrix:

$$E(:,:,r) = \begin{bmatrix} E(1,1,r) & E(1,2,r) & \cdots & E(1,m,r) \\ E(2,1,r) & E(2,2,r) & \cdots & E(2,m,r) \\ \vdots & \vdots & \vdots & \vdots \\ E(n,1,r) & E(n,2,r) & & E(n,m,r) \end{bmatrix} \quad (22)$$

Equation (23) represents green channel energies by pixels, where the value at the position (i, j) represents the green channel energy for the pixel at position (i, j):

$$E(:,:,g) = \begin{bmatrix} E(1,1,g) & E(1,2,g) & \cdots & E(1,m,g) \\ E(2,1,g) & E(2,2,g) & \cdots & E(2,m,g) \\ \vdots & \vdots & \vdots & \vdots \\ E(n,1,g) & E(n,2,g) & & E(n,m,g) \end{bmatrix} \quad (23)$$

Equation (24) represents blue channel energies by pixels, where the value at the position (i, j) represents the green channel energy for the pixel at the position (i, j). It should be noted that:

- i = {1, 2, ..., n} and
- j = {1, 2, ..., m}.

The n and m are dimensions of the image, or, to be precise, the number or rows and columns of the image.

$$E(:,:,b) = \begin{bmatrix} E(1,1,b) & E(1,2,b) & \cdots & E(1,m,b) \\ E(2,1,b) & E(2,2,b) & \cdots & E(2,m,b) \\ \vdots & \vdots & \vdots & \vdots \\ E(n,1,b) & E(n,2,b) & & E(n,m,b) \end{bmatrix} \quad (24)$$

The 3D matrix is then given with layers defined in (22–24) with the relation (25), which defines the final dimension of the matrix E:

$$E = \begin{bmatrix} E(:,:,r) \\ E(:,:,g) \\ E(:,:,b) \end{bmatrix} \quad (25)$$

where r (red), g (green), b (blue) denote the color space dimensions. Taking into account the positions of the layers in matrix (25), we can conclude that:

- r = 1,
- g = 2, and
- b = 3.

The second block in the algorithm, which is initialized first, is the accumulator (in some references expression buffer is also used for similar operation). It also contains the energies for the corresponding pixels for L frames, where the last frame is the frame in $(k-1)$ time if the current frame is designated as k.

L depends on the heuristic experience of the operator. It cannot be too large because of the reaction time. Furthermore, there is the real danger that motion will be mistaken for noise and suppressed by the averaging process. Therefore, the choice of L is limited to several frames.

The accumulator equals:

$$E_{accum}(x,y,c) = \frac{1}{L}\sum_{k=1}^{L}\left(|I_{w1}(x,y,c,k)|^2 + |I_{w2}(x,y,c,k)|^2\right) \quad (26)$$

where $E_{accum}(x, y, c)$ is the energy of the accumulator, which replaces the current frame in the background subtraction part of the algorithm. It corresponds to the current frame pixel at the spatial coordinates (x, y) for the color c. The number k corresponds to the frame number from 1 to L. The equation can be described in matrix form as well.

The energy of the current frame is calculated similarly to (21):

$$E_{current}(x,y,c) = |I_{w1}(x,y,c,k)|^2 + |I_{w2}(x,y,c,k)|^2 \quad (27)$$

All subtractions are performed by taking into account the absolute value of the results. By subtracting background and accumulator, reliable motion detection is obtained and the result is free of variations in illumination.

$$Motion\_without\_var = |E_{accum} - E| \quad (28)$$

By subtracting the current energy and background model, the motions with illumination variations are extracted:

$$Motion\_with\_var = |E_{current} - E| \quad (29)$$

The third subtraction includes the previous two differences. The motion without and with illumination variations is subtracted in the third subtraction and the result should be, in theory, only the variations:

$$Variations = |Motion\_with\_var - Motion\_without\_var| \quad (30)$$

The obtained results are scaled, because the results are of small values and difficult to visualize in the human visual system:

$$Output = Variation \cdot Scale \quad (31)$$

where $Scale > 1$, because we want magnification.

If smaller values of the matrix *Output* are needed, then *Scale* should have a value smaller than 1. In practice, this situation should never happen.

It is important to point out that the *Scale* is a constant number and not the vector of any dimension.

Finally, the resulting image is re-colored in order to emphasize the variations. The output of the described operation is seen on the screen as visualization of the calculated matrix *Output_final*:

$$Output\_final = Output \cdot [Recoloring\,matrix] \qquad (32)$$

The background model is updated by the application of the supposed variation-free version of segmented motion by having the areas of the image with motion excluded from updating and the areas without motion taken into account in the recalculation of the background model as:

$$E_{new} = \begin{cases} E \cdot \frac{(N-1)}{N} + \left(|I_{w1}(x,y,c,k)|^2 \\ + |I_{w2}(x,y,c,k)|^2\right) \cdot \frac{1}{N}, & \text{for pixels without motion} \\ E, & \ldots \text{pixels with motion} \end{cases} \qquad (33)$$

where:

- $\frac{N-1}{N}$ is the weight of the old energy matrix in the new energy matrix, and
- $\frac{1}{N}$ is the weight of current's frame energy.

These two weights are used to refresh energies for the corresponding pixels without motion (background pixels):

The energy of the background model is not changed for the corresponding pixels with motion (foreground pixels). The choice which pixel is the motion pixel, automatic motion detection is used.

The constant N is the length of the accumulator used for calculation of the energy and variations suppression. The matrix E has the same dimensions as the matrix of the wavelet coefficients at the same level of decomposition.

The quality of the final element in the proposed algorithm is very subjective, because humans have to see the visualized resulting image on the screen. It is even harder if the output unit is a printer.

The output visualization should be adjusted to the screen characteristics. It is custom to adjust the output image according to a standard output screen and not to user's specific screen.

In theory, as shown in Sect. 2.2, this procedure should eliminate noise and illumination variations.

The parallel computing should segment motion with variations.

The final result should be the extraction of illumination variations.

In order to confirm the theoretical consideration, experiments were performed and results presented in the following section.

4 Results

This section is separated in two subsections:

- experimental setup, which describes in short the used equipment and software, and

- results and the math used for programming, where the results are presented in figures.

The software and hardware used in the experiments are mentioned in the first subsection. This is important, because equipment can greatly influence the results.

The second subsection presents results in form of obtained images. Furthermore, mathematical equations used to program the computer are also given and explained.

4.1 Experimental Setup

The experimental settings for hardware and software are as follows:

- the input image is in RGB color space with a resolution of 576 × 720 pixels,
- the input image is obtained by the analog camera (namely, Samsung SDC-410),
- frame grabber card PlayTVMobile PCMCIA Type II TV Turner Card, PixelView PV-A510C(FR) is used to connect analog data from the analog camera with the digital computer,
- test configuration includes Intel Core 2 Duo CPU, T7300 @ 2.00 GHz, 1.18 GHz and 0.99 GB RAM, HDD 140 GB,
- operation system is MS Windows XP, Service Pack 3,
- application software is: Matlab R2007a
- toolboxes used in Matlab are:
  - Wavelet Toolbox 4.0,
  - Image Acquisition Toolbox 2.1, and
  - Image Processing Toolbox 5.4.

Several scenes and sequences were used in the experiments to insure more generality.

It has to be pointed out that presented images are not the only images, but just the chosen examples from the research results.

In order to experiment with the same images and obtain better results, the video sequence is recorded via Matlab in a Matlab-compatible data format. The analog camera was connected to the computer via frame grabber. The situation was real laboratory. The illumination was artificial. Recorded data were used for research in order to find a better solution for the implementation issues.

Figure 4 shows an example of the input image (randomly chosen from the recorded video sequence).

Fig. 4 Image number 101 in the video sequence, an example of input image

4.2 Results and the Math Used for Programming

If Fig. 4 is subtracted from the reference and thresholded in such a way that the lowest 10 % of image difference pixels are replaced with zeros, we obtain Fig. 5. The size of images does not match, because we presented wavelet approximation of the sourced image (which means reduction by 2 in dimensions). Figure 5a can be expressed as (34):

$$Figure = \begin{cases} image\_difference & for \ |image\_difference| \geq 0.1 \\ 0 & for \ |image\_difference| < 0.1 \end{cases} \quad (34)$$

where 0.1 is the threshold value.

To see the difference between color channels, R, G and B channels are presented as a matrix from the left to the right:

$$Figure = [R\ channel \quad G\ channel \quad B\ channel] \quad (35)$$

where 576 × 720 is the size of every channel.

So, the spatial coordinates are repeated three times, each time representing different color channels. It can be seen that the worst performance is obtained for the B channel. The channels are represented as intensity maps (gray-scale image) in Fig. 5a.

Figure 5b is obtained by taking all values from Fig. 5a (thresholded image difference), which are different from zero as whites and other pixels as blacks:

$$Figure = \begin{cases} 1 & for \ |image\_difference| \geq 0.1 \\ 0 & for \ |image\_difference| < 0.1 \end{cases} \quad (36)$$

where 0.1 is the threshold value used to obtain the results.

This is an advantage if we need a logic map (1 and 0) for further higher vision applications.

Furthermore, it is possible that not all lower-energy areas with motion can be easily seen in Fig. 5a and can be seen in Fig. 5b. Figure 5c is obtained the same way as Fig. 5a, but the threshold is only 1 %:

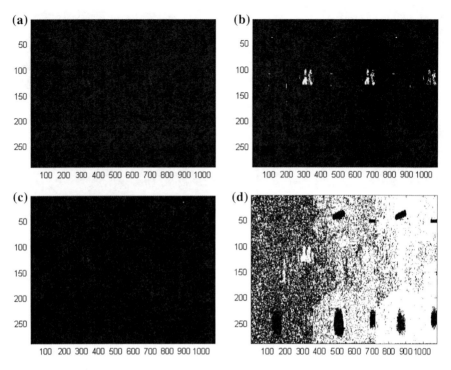

Fig. 5 Image difference (segmented motion) in *three color channels*: (**a**) each channel is represented with *gray* intensity map, (**b**) each channel is represented as *black* and *white*, (**c**) Fig. 5a with 1 % threshold, (**d**) Fig. 5c in *black* and *white* version

$$Figure = \begin{cases} image\ difference & for\ |image\_difference| \geq 0.01 \\ 0 & for\ |image\_difference| < 0.01 \end{cases} \quad (37)$$

Figure 5d is calculated with the expression:

$$Figure = \begin{cases} 1 & for\ |image\_difference| \geq 0.01 \\ 0 & for\ |image\_difference| < 0.01 \end{cases} \quad (38)$$

Figure 5d is obtained by a 1 % threshold, which is risky. It means that only 1 % of the color value can be considered black (zero value) and every difference higher than 0.01 is considered to be the actual difference and not noise.

As expected, Fig. 5d is greatly influenced by noise, which is falsely interpreted as motion and segmented into output.

It has to be pointed out that the used Matlab version uses different packaging of the matrix than newer versions. Therefore Eqs. (39–41) can be discarded if newer versions are used. Namely, the newer versions recognize RGB data and pack them in a 3D matrix, while older versions do not. The older versions (such as we used) packs 2D image data in a 2D matrix and the color data are put into the extension in number of columns by three times.

Visualization of Global Illumination Variations

Fig. 6 Color representation of image difference if color channels can only be 1 or 0

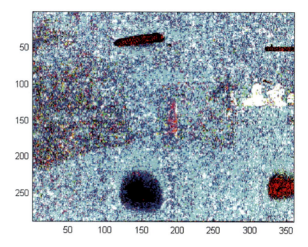

If we reconstruct wavelet approximations from R, G and B channels into a color image by the expressions:

$$Figure\ (R) = image\_difference\ (1:288, 1:360) \quad (39)$$

$$Figure\ (G) = image\_difference\ (1:288, 361:720) \quad (40)$$

$$Figure\ (B) = image\_difference\ (1:288, 721:1080) \quad (41)$$

Finally, Fig. 6 is expressed with:

$$Figure\_out = \begin{cases} 1 & if\ Figure \geq 0.1 \\ 0 & if\ Figure < 0.1 \end{cases} \quad (42)$$

which means that the color channel can be 0 or 1. If the value of some matrix element in *Figure* is less than 0.1, the *Figure_out* will be 0 (black). If the value of some matrix elements is equal or higher than 0.1, the *Figure_out* will be 1 (white). The matrix *Figure_out* is actually a logic map. Figure 6 exhibits strong noise and

Fig. 7 210th frame in the same experimantal sequence

Fig. 8 Image difference (segmented motion) in *three color channels*: (**a**) each channel is represented with *gray* intensity map, (**b**) results for recalculation of Eqs. (33–35), (**c**) *black* and *white* version of (**b**)

Fig. 9 22th frame of the training sequence

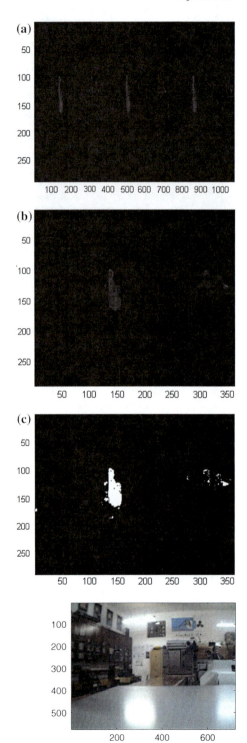

Visualization of Global Illumination Variations

Fig. 10 Image difference (segmented motion) in *three color channels*: (**a**) each channel is represented with *gray* intensity map, (**b**) results for recalculation of Eqs. (33–35), (**c**) *black* and *white* version of (**b**)

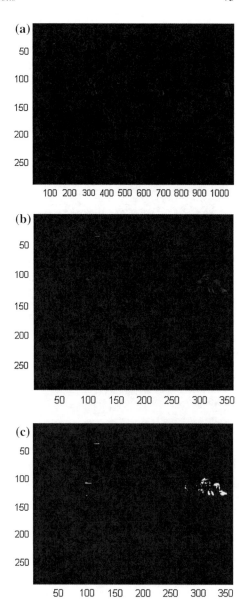

artifacts, which disables motion segmentation and pattern recognition or any other higher vision application.

To illustrate that results from Fig. 5 are not a coincidence, we will use the important expression used for Fig. 5 on the Fig. 7. Figure 7 is the 210th frame in the same video sequence. The results are presented in Fig. 8.

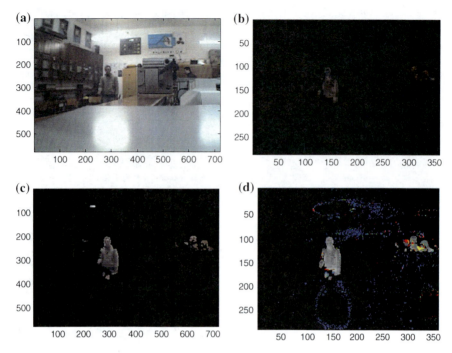

Fig. 11 (a) Frame 182 in the video sequence, (b) difference in wavelet approximations of the start frame and frame 182, (c) segmented motion, (d) difference of (c) and (b) for extraction of variations

Frame 210 is a part of the operational segment of the video sequence. However, it is important to see what is happening with the training sequence as well. Figure 9 shows frame 22 in the training sequence. Figure 10 presents the same operation as Fig. 8 for example frame i of the training sequence (frame 22).

Figure 11a shows frame 182 in the analyzed video sequence.

Figure 11b shows a different image obtained by subtraction of the start frame and frame 182 in the wavelet approximation domain. It can be seen that illumination variations have a great impact on the motion mask.

Figure 11c shows the segmented motion by the difference of the accumulator and the background model consisting of wavelet energies as explained in Sect. 3.

Figure 11d shows extracted variations in color intensities for image 182. It can be seen that areas of segmented motion are also influenced by the variations.

Examples of the outputs for the developed algorithm are shown in Fig. 12. The figure is organized as screen shots. Each shoot consists of four parts (Matlab windows)—called figures. Window called "Fig. 1" is the current input frame. Window called "Fig. 3" shows an image difference of approximations. Window "Fig. 5" shows the segmented motion with suppressed variations, and window

Visualization of Global Illumination Variations

(a)
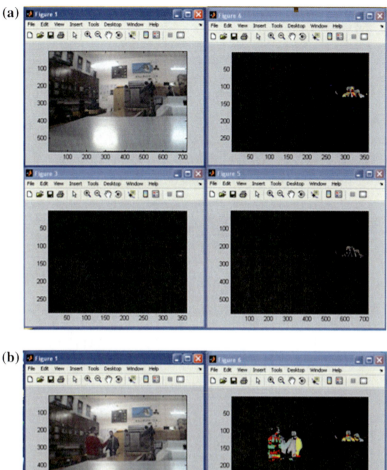

(b)

Fig. 12 Screen shots for different frames in the presented video sequence

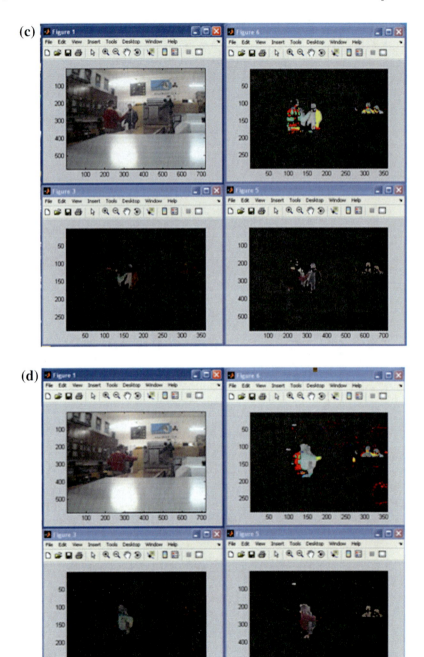

Fig. 12 continued

Visualization of Global Illumination Variations 79

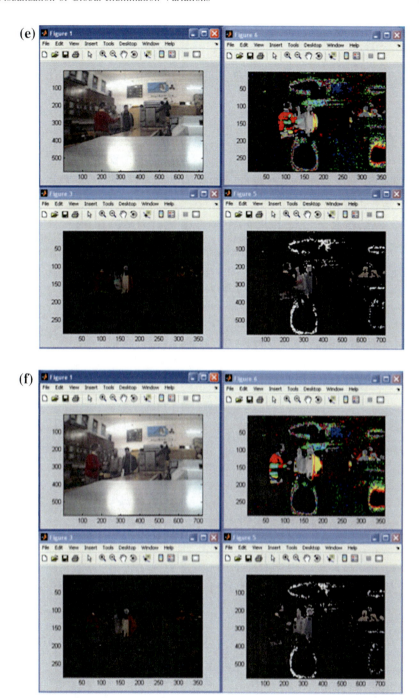

Fig. 12 continued

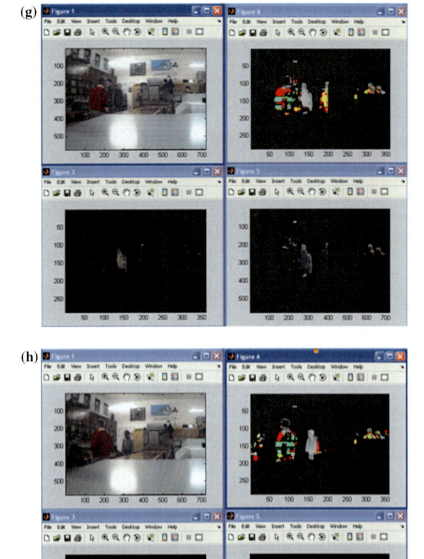

Fig. 12 continued

Visualization of Global Illumination Variations

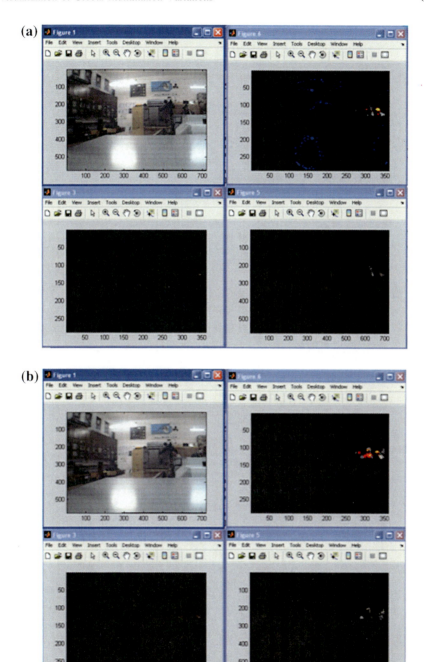

Fig. 13 Results presented as screen shots

(c)

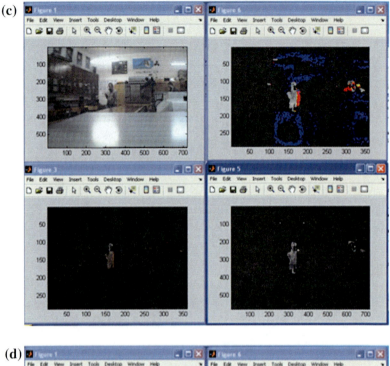

(d)

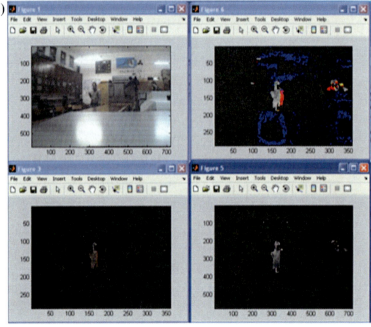

Fig. 13 continued

Visualization of Global Illumination Variations 83

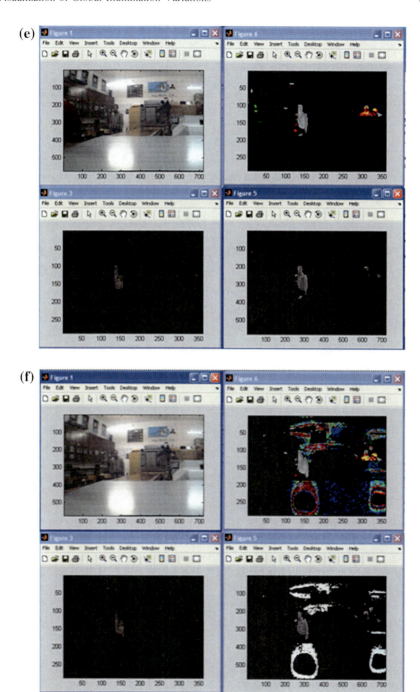

Fig. 13 continued

Fig. 13 continued

called "Fig. 6" shows the extracted variations in correlation with segmented motion.

As can be seen from Fig. 12, variations occur in some, but not in every frame. However, it is difficult to establish which one has variations in the absence of a variation-free background model.

The background model, as well as the calculations of differences and accumulator, can be performed in the wavelet domain and with energy of the wavelet coefficients as proposed in Sect. 3. Both ways present a good basis for visualization of illumination variations.

The difference between Figs. 12 and 13 is that Fig. 13 is obtained by energy calculations. "Figure 6" should be a pure variation obtained by the energy solution in the wavelet domain. "Figure 5" should be the segmented motion without variation compensation.

Figure 14 shows results of the comparison in frame 500 for wavelets methods with and without energy.

It can be concluded that the energy suppressed the false motion image from the right side of the current frame. It was possible, because there was no motion in that place in the training sequence. However, the left man is less visible than if wavelet approximation was used.

Visualization of Global Illumination Variations

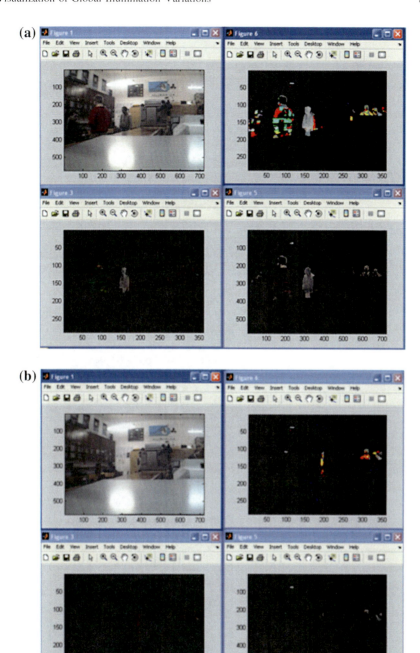

Fig. 14 (a) Screen shot for the frame 500 without energy, (b) frame 1 with energy, (c) frame 500 with energy, (d) segmented motion in energy model, (e) segmented motion without energy

Fig. 14 continued

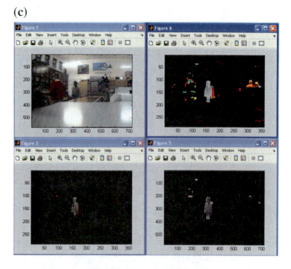

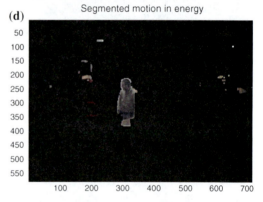

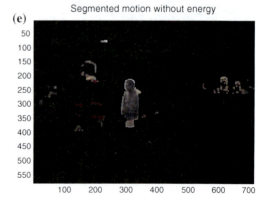

Visualization of Global Illumination Variations 87

Fig. 15 Variations detected by: (a) the energy model, (b) the non-energy model

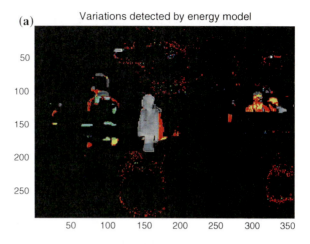

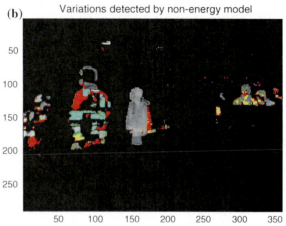

Fig. 16 Difference in segmented motion by the energy and non-energy wavelet model

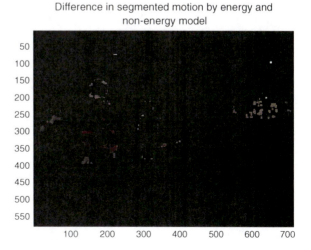

Variations detected by the energy model are shown in the Fig. 15a. It is obvious that the border left person is not segmented correctly.

Figure 15b shows the results for the same frame by the non-energy model. The border left person is segmented better than by the energy model. Both energy and non-energy model are based on wavelets.

The difference between motions segmented by the wavelet energy and non-energy method is shown for the frame in Fig. 16. If detections are the same, the resulting image should be totally black (zero-matrix). However, this is not the case.

5 Conclusions

The usual approach in image/video processing is to try to suppress illumination variations. In this chapter, variations are detected in order to emphasize them. Pure variations were attempted to be extracted. Variations are present in the foreground and the background. However, for motion segmentation purposes the variations in the background are relevant only if they influence the detection of motions (if variations exceed the threshold). If the threshold is fixed, the variations can be greatly reduced by its higher value. The disadvantage of the fixed higher threshold is in the over-segmentation of the output motion mask. In some cases, the motion would not be detected at all. Therefore, the threshold must be adaptable. In the experiments presented, the threshold is defined as a percentage of energy. The percentage is fixed, but energy is variable. Figure 5 shows influence of the threshold (1 and 10 %) on the output. The threshold is carried out by Eq. (34) or modified Eq. (34) by the change in threshold value.

The visualization of illumination variations can be realized by differencing the reference model (variation-free image) and the current frame (variations-poisoned image). After such extraction, the variations can be re-colored or emphasized in some other way. In this chapter, the RGB color space was used.

Further research should include techniques for threshold selection and techniques for human-adapted visualization of variations. Furthermore, a way to safely remove variations in illumination should be investigated.

References

1. Zhang, T., Tang, Y.Y., Fang, B., Shang, Z., Liu, X.: Face recognition under varying illumination using gradientfaces. IEEE Trans Imag Process **18**, 2599–2606 (2009)
2. Choi, M., Kim, G., Choi, H.: Robust character region extraction against camera motion and illumination variation. Proceedings of the 7th WSEAS international conference on computational intelligence, Man-machine systems and cybernetics, pp. 161–164 (2008)
3. Porter, R., Fraser, A.M., Hush, D.: Wide-area motion imagery. IEEE Sig Process Mag **27**, 56–65 (2010)

4. Vujović, I.: Suppressing illumination variations in motion detection by wavelet transform. Ph.D. Thesis, Faculty of Electrical Engineering, Mechanical Engineering and Naval Architecture, University of Split, Split (2011)
5. Wong, P.C., Bergeron, R.D.: Multiresolution multidimensional wavelet brushing. Proceedings of IEEE visualization, pp. 171–178 (1996)
6. Stauffer, C., Grimson, W.E.L.: Adaptive background mixture models for real-time tracking. Proceedings of IEEE computer society conference computer vision and pattern recognition, Ft. Collins, USA, vol. 2, pp. 246–252 June 1999
7. Theiler, J.: Quantitative comparison of quadratic covariance-based anomalous change detectors. Appl. Opt. **47**, F12–F26 (2008)
8. Porter, R., Harvey, N., Theiler, J.: A change detection approach to moving object detection in low fame-rate video. Proceedings of SPIE, Orlando, USA, 73410S(8) (2009)
9. Rosin, P., Ioannidis, E.: Evaluation of global image thresholding for change detection. Pattern Recogn. Lett. **24**, 2345–2356 (2003)
10. Christopher, H., Walnut, D.F.: Fundamental Papers in Wavelet Theory. Princeton University Press, London (2006)
11. Jansen, M., Oonincx, P.: Second Generation Wavelets and Applications. Springer-Verlag, London (2005)
12. Mallat, S.: A Wavelet Tour of Signal Processing. Academic Press, New York (2009)
13. Vujović, I., Šoda, J., Kuzmanić, I.: Cutting-edge mathematical tools in processing and analysis of signals in marine and navy. Trans Marit Sci **1**, 35–46 (2012)
14. Tolba, M.F., Bahgat, S.F., Al-Berry, M.N.: Wavelet-enhanced detection of low contrast objects moving in environments with varying illumination. Int J Intell Coop Inf Syst **5**, 395–412 (2005)
15. Tolba, MF., Bahgat, S.F., Al-Berry, M.N.: A fast and reliable memory-based frame-differencing technique for moving object detection. Proceedings 14th international conference on computer: theory and applications, Alexandria, Egypt (2004)
16. Matlab homepage. Available at: http://www.mathworks.com
17. Selesnick, I.W., Li, K.Y.: Programs for 3-D oriented wavelet transforms and examples. Available at: http://taco.poly.edu/WaveletSoftware (2003)
18. Führ, H., Demaret, L., Friedrich, F.: Beyond wavelets: new image representation paradigms. In: Barni, M. (ed.) Document and Image Compression. CRC Press, London (2006)
19. Melchior, P., Meneghetti, M., Bartelmann, M.: Reliable shapelet image analysis. Astron. Astrophys. **463**, 1215–1225 (2007)
20. Pennec, E., Mallat, S.: Sparse geometric image representations with bandelets. IEEE Trans Imag Process **14**, 423–438 (2005)
21. Wu, B., Nevatia, R.: Detection of multiple, partially occluded humans in a single image by bayesian combination of edgelet part detectors. 10th IEEE international conference on computer vision, Beijing, China, October 17–20, pp. 90–97 (2005)
22. Candés, E., Demanet, L., Donoho, D., Ying, L.: Fast discrete curvelet transforms. Multiscale Model Simul **5**, 861–899 (2006)
23. Amer, A.: Memory-based spatio-temporal real-time object segmentation for video surveillance. Proceedings of the conference on real-time imaging VII, Santa Clara, January 22–23, vol. 5012, pp. 10–21 (2003)
24. Zhichao, L., Joo, E.M.: Face recognition under varying illumination. In: Er, M.J. (ed.) New Trends in Technologies: Control, Management, Computational Intelligence and Network Systems. InTech, Rijeka (2010)
25. Dorf, R.C.: The Electrical Engineering Handbook. CRC Press LLC, Boca Raton (2000)
26. Vujović, I., Kuzmanić, I., Beroš, S.M., Šoda, J.: Choosing wavelet pairs in suppression of illumination variations for port surveillance, Proceedings Electronics in Marine ELMAR 2011, Zadar, Croatia, September 14–16, pp. 75–78 (2011)

Evaluation of Fatigue Behavior of SAE 9254 Steel Suspension Springs Manufactured by Two Different Processes: Hot and Cold Winding

Carolina Sayuri Hattori, Antonio Augusto Couto, Jan Vatavuk, Nelson Batista de Lima and Danieli Aparecida Pereira Reis

Abstract The fatigue resistance is a property that exerts a strong influence on the suspension spring performance in vehicles. The choice of SAE 9254 steel was due to its wide use in the manufacture of these springs and their fatigue properties and toughness. The manufacture of SAE 9254 steel springs has been made by the hot winding process and the heat treatment by conventional quenching and tempering or by cold winding process and induction hardening and tempering. The shot peening induced a compressive residual stress which increased the fatigue life of the SAE 9254 steel. The residual stress profile from the surface of springs showed a peak in the values of the compressive stress for both manufacturing processes. The maximum residual stress in the cold processed spring was higher than the hot processed spring and maintained much higher values along the thickness of the spring from the surface, resulting from manufacturing processes. The fatigue cracking of the springs, without shot peening, started by torsional fatigue process, with typical macroscopic propagation. The fracture surface showed stretch marks with high plastic deformation.

1 Introduction

The main components of automobile suspension systems are: springs, shock absorbers and stabilizing rods (installed in the chassis of the vehicle). Other components such as pivots (or peripheral pins) and brackets (or suspension tray) can be also considered to be part of the support structure. Marked efforts are ongoing to develop high performance spring steels to address demands of the automobile industry to reduce costs and weight. Since fatigue strength and

---

C. S. Hattori · A. A. Couto (✉) · J. Vatavuk · N. B. de Lima · D. A. P. Reis
IPEN-CNEN/SP and Mackenzie Presbyterian University, Rua da Consolação,
930, São Paulo, SP 01302-907, Brazil
e-mail: acouto@ipen.br

resistance to slackening are important properties of spring steels, one of the objectives of this study was to improve these properties through heat treatment, addition of micro alloying elements and/or mechanical treatments such as shot peening [1, 2].

The 92xx series SAE steels are listed, as per standards, as carbon steels with 1.2–2.0 % Si, 0.55 % Cr and/or 0.55 % Mn. The SAE 9254 steel in the form of drawn or wire rod is widely used in the manufacture of suspension springs as it has excellent properties in terms of mechanical strength, fatigue strength and toughness. The processes to produce suspension springs differ in terms of the heat treatment given to the spring to attain the final properties and desired characteristics. The first process that is widely used in industry is hot winding in which the heat treatment of the spring consists of conventional quenching and tempering. In the second manufacturing process, the heat treatment is carried out on the wire rod at the beginning of the process. The wire rod is induction hardened on-line and the spring is subsequently cold wound.

The production processes of suspension springs differ in the heat treatment for obtaining the final properties and characteristics needed for its application. One method widely used in industry is the hot winding process, where the spring is submitted to heat treatment of quenching and tempering by the conventional process. In another process, the heat treatment is performed in the wire rod at the beginning of the process. This induction hardening and tempering on the wire rod is done online and the spring is cold winded later. Therefore, it is necessary to analyze the effect of both manufacturing spring suspension, aiming at identifying the process that has the best performance with regard to mechanical properties and microstructure. It is also important to check the residual stress in the springs with and without shot peening by X-ray diffraction analysis.

The aim of this study was to compare the mechanical properties (fatigue and hardness), microstructure (optical microscopy) and fracture surfaces of SAE 9254 steel suspension springs, with and without shot peening, and manufactured by the hot winding process followed by conventional quenching and tempering as well as by the cold process of induction hardening. The residual stress induced in the springs by the two processes, with and without shot peening was determined using X-ray diffraction analysis.

2 Materials and Methods

In this study, SAE 9254 steel with a composition as shown in Table 1 was used to manufacture suspension springs. This steel, received as drawn rods and wire rods without heat treatment, was then sent to manufacture the springs. For the hot winding process were used drawn bars of 12.07 mm in diameter. The springs were manufactured by the hot process, consisting of quenching and tempering, as well as by the cold process in which they were induction hardened. In the hot winding process, drawn bars were used as the starting material. The bars were austenitized at 880 °C, hot wound and oil quenched. After quenching, the springs were tempered at 400 °C.

For the cold winding process, wire rods of 12.07 mm in diameter were used. In the cold winding process the wire rod was first pickled mechanically and then inspected using the eddy current technique. After this stage, the wire was induction hardened at ~950 °C. The tempering was carried out at 470 °C, also using an induction coil. The wire was then inspected once more for cross-sectional surface defects using the eddy current technique, prior to winding for the subsequent stages of the process. The quenched and tempered wire was wound at room temperature to the required sizes and then stress relieved at 180 °C.

The springs were also shot peened. The impact of shots produces a layer having compressive residual stresses on the workpiece surface treated. Each shot acts like a small hammer causing plastic deformation on the surface, tending to elongate. However, this tendency is impeded by deeper inner layers which are elastically deformed during the impact. As a result of the interaction of these two layers, superficial and internal, compressive residual stresses originate in plastically deformed layers. The most of the fatigue failures occur on the surface or in near areas. The tensile stresses are responsible for crack initiation. These cracks have difficulties to nucleate or propagate in a field of compressive stresses. The compressive residual stresses induced by shot peening produce a significant increase in fatigue life of the component.

There are two very important parameters to control the shot peening process, Almen [3] intensity and coverage. The intensity of shot peening is given by the particle impact force on the workpiece. The higher the density and particle hardness the greater the shot peening intensity. Higher particle velocity causes a greater impact on the workpiece. The intensity is directly related to the kinetic energy of the particles. The Almen test is the standard measurement system to control the intensity of shot peening. This technique is used to estimate the intensity of the process for some operating parameters and assumes that equal deformations in standard thin plates correspond to applications with equal intensities. The Almen intensity is governed by the following process parameters: size and hardness of the shot; turbine rotation; jet speed and angle of the jet.

The cover is a parameter that estimates how complete an area subjected to shot peening was covered by indentations created by the impacts of shots. In practice, a measurement of coverage of 98 % or greater is considered as 100 % and the time used is defined as saturation time. Shot peening executed in shorter times of saturation are not common. This is due to the fact that small areas not affected by the indentations could act as initiation points of failure, so that would damage mechanical properties. For this reason, the coverage is expressed in percentage and

Table 1 Nominal chemical composition of SAE 9254 steel (wt %)

C	Mn	Si	Cr
0.50–0.60	0.60–0.80	1.20–1.60	0.60–0.80

as multiple of the exposure time required for saturation. The parameters that directly affect the coverage are: exposure time, size of shots and mean flow of shots.

In the shot peening process, 0.4–0.8 mm diameter 'cut wire' type of shots with hardness in the range of 610–670 HV were used. The shot peening intensity was measured on a type A strip using the Almen test and the minimum and maximum values were 0.25 and 0.4 mm, respectively (curvature or deformation of the sheet). The shot peening duration was 6–10 s and the coverage was 99.9 %.

Conventional metallographic procedures consisting of cutting, grinding, polishing, chemical etching with 2 % Nital and observation in a BX60 Olympus microscope coupled to an image analysis system were used to study the springs manufactured by the two processes. A Wolpert apparatus was used to determine the Rockwell C hardness of the spring samples. Five measurements were made on each sample. The residual stress analysis of the springs manufactured by the two processes was carried out using a Rigaku Dmax X-ray diffractometer and 20 mm samples that were cut from springs made by the two processes. The method of X-ray diffraction for residual stress detection is based on changing the interplanar distances of the crystalline structure of the material and is measured by the angular position of the diffracted X-rays.

The fatigue properties of springs manufactured by the hot and cold processes, as well as with and without shot peening were determined. The fatigue tests were carried out as per the specifications of the different manufacturers to determine the quality of their springs and the effect of shot peening on fatigue life. To carry out the fatigue tests on hot processed springs the specified minimum and maximum loads were 3,500 and 3,721 N, respectively. The minimum and maximum heights of the tests were 126.2 and 255.5 mm respectively and the free height was approximately 382.1 mm. In fatigue tests with springs that were cold processed the minimum and maximum loads were 2,907 and 3,057 N, respectively. To permit comparison of fatigue behavior of springs that were hot and cold processed and shot peened, the same procedure was used all tests. In these tests, the specified load was 3,000 N, and minimum and maximum heights were 192 and 222 mm, respectively. During the tests, the fatigue stroke of the springs was measured. Figure 1 shows the fatigue test machine used in suspension springs. The fracture surfaces of the springs that were fatigue tested were first observed in a stereo microscope and then in a Jeol JSM—6510 model scanning electron microscope.

3 Results and Discussion

A typical microstructure of as received SAE 9254 steel without heat treatment is shown in Fig. 2 and it consists of coarse pearlite and ferrite. The microstructure of a hot processed spring shown in Fig. 3 is typical for the given quench and temper heat treatment. Due to the austenitization treatment given at 820 °C, partial dissolution of carbon and the carbides was homogeneous, resulting in a fine

Fig. 1 Machine used in fatigue tests of the suspension springs

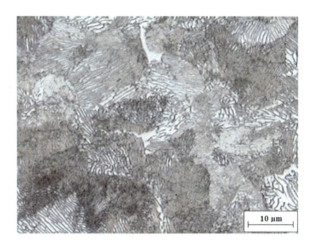

Fig. 2 Microstructure of the samples as received

martensite with uniformly distributed carbides. Figure 4 shows the spring that was induction hardened at around 950 °C. In this micrograph it is possible to note that in spite of the high heating rates, a feature of the induction hardening process, the microstructure is as homogeneous as that of a spring given a conventional quench (conventionally hardened). The micrographs in Figs. 3 and 4 do not reveal marked differences between samples that were hot processed followed by a conventional quench and samples that were cold processed and induction hardened.

The hardness results of SAE 9254 steel samples that were not heat treated, springs that were hot processed followed by conventional quenching and tempering and springs that were cold processed followed by induction hardening were 32.8, 51.8 and 53.6 HR_c, respectively. Note that the hardness values are in agreement with the heat treatments that were given to the springs. The slightly higher hardness of the spring that was cold processed, quenched and induction hardened, compared to that of the spring that was hot processed followed by

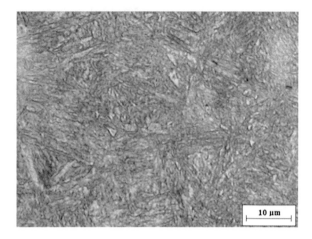

Fig. 3 Microstructure of sample that was hot processed, quenched and tempered

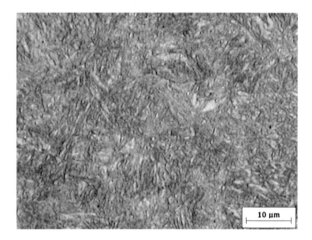

Fig. 4 Microstructure of sample that was cold processed and induction hardened

conventional quenching and tempering is not significant. Figures 5 and 6 show details of the surfaces of samples that were cold processed followed by induction hardening, without and with shot peening, respectively. Note that the surface of the sample in Fig. 6 shows irregularities caused by shot peening.

Table 2 shows the results of fatigue tests carried out by the manufacturers under conditions specified for springs that were hot processed. Table 3 shows the results of fatigue tests carried out by the manufacturers under conditions specified for springs that were cold processed. The shot peened springs that were processed by the hot process and the cold process attained their respective values without failure as specified by the manufacturers. The specification for the hot processed springs is 450,000 cycles whereas that for the cold processed spring is 1,000,000 cycles. Since the manufacturer of the hot processed springs interrupts the test upon attaining 450,000 cycles, it has not been possible to determine if these springs

Fig. 5 Detail of microstructure of surface of induction hardened sample without shot peening

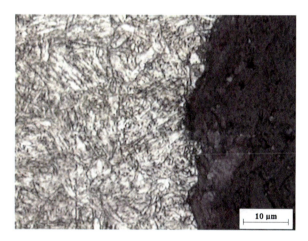

Fig. 6 Detail of microstructure of surface of sample that was induction hardened and shot peened

would have withstood 1,000,000 cycles like the cold processed springs. It is worth pointing out that the loading in fatigue tests for hot processed springs by the manufacturer is more rigid. For the hot processed springs the maximum and minimum were 285 and 114 mm respectively, whereas the maximum and minimum for the cold processed springs were 255 and 126 mm, respectively.

There is a sharp drop in the number of cycles to rupture of springs manufactured by both processes and which were not shot peened. Again, a more substantiated comparison is difficult because of the differing test conditions and specifications for springs manufactured by the two processes. However, it can be said that springs that were manufactured by the two processes and shot peened met the specifications of the respective manufacturers. As mentioned earlier, an independent laboratory carried out identical fatigue tests on springs manufactured by the two processes. The two types of springs exceeded without failure the maximum specified number of cycles (1,000,000 cycles). In these tests, the heights

Table 2 Fatigue test results of springs that were hot processed followed by conventional quenching and tempering

Springs manufactured by the hot process		Number of cycles	Ruptured
With shot peening	Sample 1	450,000	No
	Sample 2	450,000	No
Without shot peening	Sample 3	119,440	Yes
	Sample 4	175,700	Yes

Table 3 Fatigue test results of springs that were cold processed and induction hardened

Springs manufactured by the cold process		Number of cycles	Ruptured
With shot peening	Sample 1	1,000,000	No
	Sample 2	1,000,000	No
Without shot peening	Sample 3	119,440	Yes

of the springs after cycling were also measured. The height of the hot processed spring was 191 mm and that of the cold processed spring, 220 mm. That is, after the test, the difference between the initial and final heights of both springs was minimum, indicating qualification of springs made by both processes. Since the springs did not fail and the results were similar, it is difficult to rate which type of spring is better.

Tables 4 and 5 show the results of residual stress measurements carried out by X-ray diffraction analysis on springs that were cold and hot processed, with and without shot peening. Table 4 shows the results of residual stress measurements carried out on concave, convex and neutral side surfaces of the springs. Due to operational problems it was not possible to determine the residual stress in the neutral position of the spring that was cold processed and shot peened. The residual stress measurements on the surface of the springs hot and cold processed, with and without shot peening, showed compression. The exception was the neutral position of the spring hot processed and not shot peened. This latter showed low tension and the spring that was cold processed and in the convex position showed high tension. The residual stress on surfaces of springs manufactured by both processes and later shot peened was high compression. Since the hot processed springs were austenitized, hot wound, quenched and tempered (stress relieved), the residual stress on samples that were not shot peened was much lower than that on samples that were cold processed, as winding (cold deformation) was carried out after induction hardening. This was observed in the cold processed spring, without shot peening, which showed high tension on the convex surface.

Table 5 presents the residual stress profile on the surface of springs that were hot and cold processed. Figures 7 and 8 show the residual stress curves as a function of depth from the surface of springs manufactured by the two processes. The curves in these figures and data in Table 5 indicate that the residual stress in compression increased from the surface down to a certain depth and thereafter decreased continuously across the section, the overall thickness being

Evaluation of Fatigue Behavior

Table 4 Residual stress measured by X-ray diffraction on springs that were hot and cold processed

Part of the sample	Manufactured by the hot process		Manufactured by the cold process	
	Stress (MPa)		Stress (MPa)	
	Without shot peening	With shot peening	Without shot peening	With shot peening
Concave	−36.0	−462	−451	−515
Convex	−35.5	Table 5	+374	Table 5
Neutral (Side)	+40.5	−436	−112	*

(*) Measurement in this position was not possible

Table 5 Residual stress variation from the surface of springs manufactured by the hot and cold process as measured by X-ray diffraction

Hot process		Cold process	
Depth (μm)	Stress (MPa)	Depth (μm)	Stress (MPa)
0	−492	0	−541
82	−683	52	−680
99	−731	86	−752
130	−695	128	−831
157	−594	158	−840
186	−518	183	−793
234	−231	228	−777
251	−199	253	−657
308	−101	310	−501
358	−61	358	−477
423	−32	413	−426

approximately 400 μm. Table 4 also shows that in hot processed springs, the maximum stress (compression) varied from −731 MPa at a depth of 99 μm to −32 MPa at a depth of 423 μm. The maximum stress in the cold processed spring was −840 MPa at a depth of 158 μm and remained at −426 MPa at the lowest measured depth. That is, the compressive stress remained high in the cold processed spring. This fact could be of relevance while comparing the two spring manufacturing processes.

Fracture surfaces of springs that were hot and cold processed without shot peening (only ruptured springs) as seen with a stereo microscope and with a scanning electron microscope are shown in Figs. 9, 10, 11, 12, 13, 14, 15, 16, 17 and 18. Examination of the fracture surface of the hot processed spring (Fig. 9) showed that propagation by torsion fatigue occurred from the inner side, yielding to the stress in the component at maximum load. The red arrow indicates the fracture region where the propagation mechanism was torsional fatigue. Figure 10 shows at higher magnification the region indicated by the red arrow in Fig. 9. This fractography reveals the propagation mechanism with fatigue grooves attributable

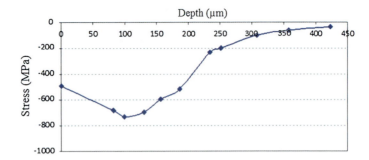

Fig. 7 Residual stress as a function of depth from the surface of spring that was hot processed

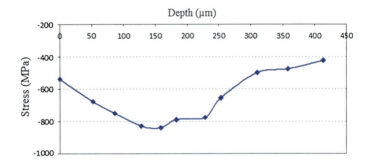

Fig. 8 Residual stress as a function of depth from the surface of spring that was cold processed

Fig. 9 Fracture surface of hot processed spring, as seen in a stereo microscope

Evaluation of Fatigue Behavior

Fig. 10 Detail of fracture surface indicated by the *red arrow* in Fig. 8 of a hot processed spring, as seen in a scanning electron microscope

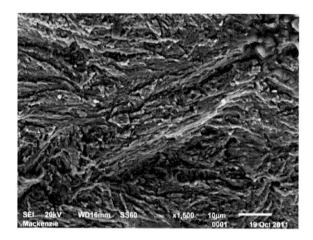

Fig. 11 Fracture surface detail of region indicated by the *blue arrow* in Fig. 8 of a hot processed spring, as seen in a scanning electron microscope

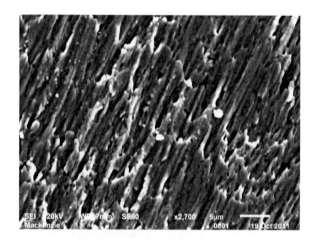

Fig. 12 Fracture surface detail of region indicated by the *orange arrow* in Fig. 8 of a hot processed spring, as seen in a scanning electron microscope

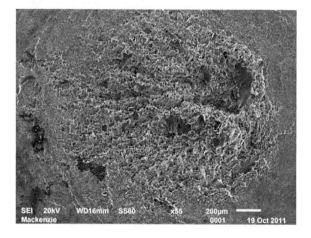

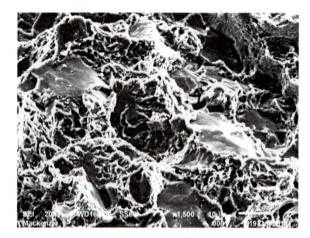

Fig. 13 Fracture surface detail of region indicated by the *orange arrow* in Fig. 8 of a hot processed spring, as seen in a scanning electron microscope

Fig. 14 Fracture surface of a cold processed spring, as seen in a stereo microscope

to the high hardness, complex distribution of crystallites besides the high internal stress state.

Figure 11 shows at a higher magnification the region shown with a blue arrow in Fig. 9. This fracture surface shows tear dimples, typical of ductile torsion fracture and the macroscopic path perpendicular to the spring axis (axis of the spring) corroborates this statement. To begin with it can be inferred that the crack was initiated by a torsional fatigue process with typical macroscopic propagation that revealed striations with high plastic deformation caused by the high shear stresses. This region of the fracture corresponds to stage II fatigue. The region corresponding to stage I of the fatigue process is difficult to identify, and this is typical in the case of steels with hardness over 50 HRC. Stage III corresponds to

Evaluation of Fatigue Behavior

Fig. 15 Fracture surface of a cold processed spring, as seen in a scanning electron microscope, revealing the region where propagation started due to torsional fatigue

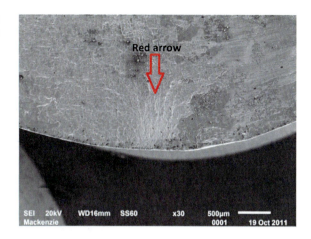

Fig. 16 Fracture surface of a cold processed spring, as seen in a scanning electron microscope, revealing the region shown by the *red arrow* in Fig. 13

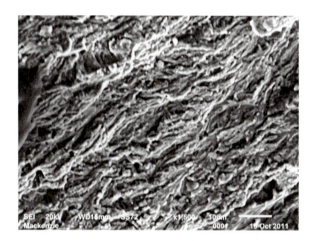

Fig. 17 Fracture surface of a cold processed spring, as seen in a scanning electron microscope, revealing the region shown by the *orange arrow* in Fig. 13

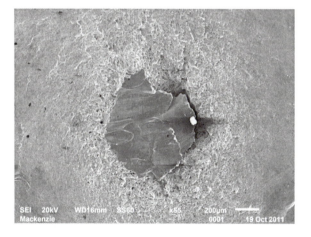

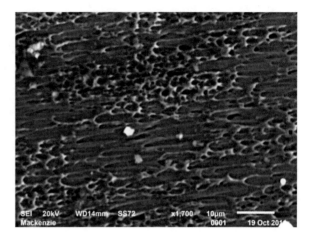

Fig. 18 Fracture surface of a cold processed spring, as seen in a scanning electron microscope, revealing the region shown by the *orange arrow* in Fig. 13

the end of propagation of the crack and in this case it occurred by a ductile torsion fracture mechanism, perpendicular to the axis and leading to final separation. This is indicated by the orange arrow in Fig. 9. This mechanism is shown in detail in Figs. 12 and 13.

Figures 14, 15, 16, 17 and 18 shows fracture surfaces of springs that were cold processed and without shot peening. On the fracture surface of the spring manufactured by the cold process, shown in Fig. 14, the region corresponding to initiation of propagation by torsional fatigue is shown with a red arrow. However, it is less accentuated, compared to that in springs manufactured by the hot process. The sample that was cold processed showed a smaller crack, resulting from fewer stress multiplier and lower toughness compared to a hot processed spring. In the latter, stage II crack propagation (red arrow in Fig. 14) was more due to higher nominal stress. Thus, it can be inferred that the hardened martensite is less tough compared to unhardened martensite (hot process). Stage II shows signs of striations (Figs. 15 and 16) associated with extreme shearing caused by the nature of the load. The end of stage III (orange arrow in Fig. 14) is shown in Figs. 17 and 18. A sheared region can be seen in Fig. 17, probably caused by cutting forces, which could be low due to deformation caused by load relief in the spring at final fracture. In Fig. 18 the presence of micro cavities adjacent to the sheared region (Fig. 17) is evident.

4 Conclusions

Comparison between the SAE 9254 suspension springs manufactured by the hot process followed by conventional quenching and tempering with that obtained by the cold process followed by quenching and induction hardening enabled the following preliminary conclusions:

- The hardness of the springs in the different conditions is compatible with the heat treatments and the corresponding microstructures. The microstructure of the quenched springs showed the presence of tempered martensite and carbides. In springs manufactured by both processes, optical microscopy did not reveal significant differences in microstructures.
- Shot peening induced compressive residual stress and this increased markedly the fatigue strength of SAE 9254 steel. Springs manufactured by both processes and shot peened were found to be acceptable, as per fatigue test specifications of the manufacturers.
- Use of the same fatigue test procedure on springs manufactured by the hot and cold process did not permit classification of the manufacturing process in terms of improved fatigue properties.
- The residual stress profile from the surface inwards of the shot peened spring showed a peak in the compression stress for both manufacturing processes. The maximum residual stress in the cold processed and shot peened spring was higher than that of the hot processed spring. This value remained higher across the thickness of the spring, starting from the surface.
- The fatigue cracks on springs manufactured by both processes and without shot peening initiated by a torsional fatigue process and with typical macroscopic propagation. The fracture surface revealed signs of striations with high plastic deformation, caused by forces that increased shear stresses.

Acknowledgments The authors thank the financial support through a scholarship granted by the Presbyterian Mackenzie Institute for Carolina Sayuri Hattori. The authors also acknowledge support by the ArcelorMittal, Allevard Molas do Brasil and Mubea do Brasil companies in providing materials and carrying out fatigue tests.

References

1. Lee, C.S., Lee, K.A., Li, D.M., Yoo, S.J., Nam, W.J.: Microstructural influence on fatigue properties of high-strength spring steel. Mat Sci Eng **A241**, 30–37 (1999)
2. Canaan, G.L.: Influência da Adição de Vanádio nas Propriedades Mecânicas de Aços Médio Teor de Carbono para Beneficiamento. Universidade Federal de Minas Gerais, Belo Horizonte (2007)
3. Almen, J.O., Black, P.H.: Residual Stresses and Fatigue in Metals. MacGraw Hill, Book Company, New York (1963)

Yield Criteria for Incompressible Materials in the Shear Stress Space

Vladimir A. Kolupaev, Alexandre Bolchoun and Holm Altenbach

Abstract In the theory of plasticity different yield criteria for incompressible material behavior are used. The criteria of Tresca, von Mises and Schmidt-Ishlinsky are well known and the first two are presented in the textbooks of Strength of Materials. Both Tresca and Schmidt-Ishlinsky criteria have a hexagonal symmetry and the criterion of von Mises has a rotational symmetry in the π-plane. These criteria do not distinguish between tension and compression (no strength differential effect), but numerous problems are treated in the engineering practice using these criteria. In this paper the yield criteria with hexagonal symmetry for incompressible material behavior are compared. For this purpose, their geometries in the π-plane will be presented in polar coordinates. The radii at the angles of 15° and 30° will be related to the radius at 0°. Based on these two relations, these and other known criteria will be shown in one diagram. In this diagram the extreme shapes of the yield surfaces are restricted by two criteria: the Unified Yield Criterion (UYC) and the Multiplicative Ansatz Criterion (MAC). The models with hexagonal symmetry in the π-plane for incompressible materials can be formulated in the shear stress space. For this formulation platonic, archimedean and catalan solids with orthogonal symmetry planes are used. The geometrical relations of such models in the π-plane will be depicted in the above mentioned diagram. The examination of the yield surfaces leads to the generalized

V. A. Kolupaev (✉)
Mechanik und Simueation, Deutsches Kunststoff-Institut (DKI),
Schloßgartenstr. 6 64289 Darmstadt, Germany
e-mail: VKolupaev@dki.tu-darmstadt.de

A. Bolchoun
KC Bauteilgebundenes Werkstoffverhalten, Fraunhofer-Institut für Betriebsfestigkeit
und Systemzuverlässigkeit LBF, Bartningstr. 47, 64289 Darmstadt, Germany
e-mail: alexandre.bolchoun@lbf.fraunhofer.de

H. Altenbach
Lehrstuhl für Technische Mechanik, Institut für Mechanik Fakultät für Maschinenbau
Otto-von-Guericke-Universität Magdeburg, Universitätsplatz 2, 39106 Magdeburg, Germany
e-mail: holm.altenbach@ovgu.de

criterion with two parameters. This model describes all possible convex forms with hexagonal symmetry. The proposed way to look at the yield criteria simplifies the selection of a proper criterion. The extreme solutions for the analysis of structural members can be found using these criteria.

Keywords Flow criteria · Equivalent stress · Deviatoric plane · Hexagonal symmetry · Generalization

1 Introduction

Due to the assumption of incompressibility the surface Φ which defines a yield criterion is a function of the invariants of the stress deviator (Appendix)

$$\Phi(I'_2, I'_3, \sigma_{eq}) = 0. \qquad (1)$$

In general the surface Φ has a trigonal symmetry in the π-plane. The equivalent stress σ_{eq} is equal to the yield stress at tension σ_+

$$\sigma_{eq} = \sigma_+. \qquad (2)$$

From numerous tensile and compression tests of ductile materials (mild steel, brass, nickel, copper, etc.) it follows that

$$\sigma_+ = \sigma_-. \qquad (3)$$

σ_- is the yield stress at compression [44]. Because of (3) the convex surface Φ in the case of the incompressibility

$$\Phi(I'_2, I'^2_3, \sigma_{eq}) = 0 \qquad (4)$$

has a hexagonal symmetry in the π-plane (the plane orthogonal to the hydrostatic axis in the principal stress space). In the special case if the influence of I'_3 can be neglected, a rotationally symmetrical shape occurs in the π-plane (von Mises criterion)

$$\sigma^2_{eq} = 3 I'_2. \qquad (5)$$

The criteria of Tresca [31, 32, 37]

$$\left(I'_2 - \sigma^2_{eq}\right)^2 \left(2^2 I'_2 - \sigma^2_{eq}\right) - 3^3 I'^2_3 = 0 \qquad (6)$$

and Schmidt-Ishlinsky [3, 7, 12, 15, 33, 40–42]

$$\left[\frac{3^3}{2^3} I'_3 + \frac{3^2}{2^2} I'_2 \sigma_{eq} - \sigma^3_{eq}\right] \left[\frac{3^3}{2^3} I'_3 - \frac{3^2}{2^2} I'_2 \sigma_{eq} + \sigma^3_{eq}\right] = 0 \qquad (7)$$

have found wide applications [44]. Their shape in the π-plane is regular hexagon (Fig. 1, left). Furthermore, other criteria are known, e.g. regular dodecagons of Sokolovsky [5, 30, 40] and Ishlinsky-Ivlev [16–18] in the π-plane (Fig. 1, right). New criteria can be be established using the formulation of criteria in the shear stress space (Sect. 4), but up to now no systematization has been proposed.

In this article it will be shown that the Unified Yield Criterion (UYC) is the conservative form of the surface Φ (Sect. 3). The upper limit is given by the Multiplicative Ansatz (MAC) of the Tresca and the Schmidt-Ishlinsky criterion. The UYC and the MAC are functions of the same parameter and permit generalization with one additional parameter. The resulting model with two parameters describes all possible convex forms of hexagonal symmetry in the π-plane.

The aim of this paper is to systematize various yield criteria. To achieve this, they are shown in a diagram based on the geometrical properties of surfaces in the π-plane (Sect. 2). Thereby some existing gaps in the modeling are filled.

2 Basic Relations

For the analysis of the yield criteria for incompressible material behavior their geometries are compared in the π-plane (Fig. 2). Therefore, the radii at the angles $\theta = \frac{\pi}{12}, \frac{\pi}{6}$ and $\frac{\pi}{3}$ are related to the radius at the angle $\theta = 0$:

$$h = \frac{R\left(\frac{\pi}{12}\right)}{R(0)}, \quad k = \frac{R\left(\frac{\pi}{6}\right)}{R(0)}, \quad d = \frac{R\left(\frac{\pi}{3}\right)}{R(0)}. \tag{8}$$

For the model of von Mises it follows

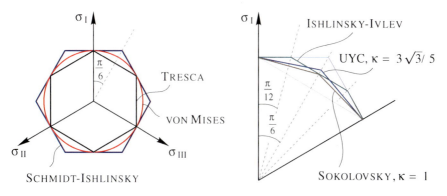

Fig. 1 Continuous surfaces Φ with hexagonal symmetry in the π-plane and the model of von Mises (5), incompressible material behavior with the property $\sigma_+ = \sigma_-$ [25]. On the right hand side an enlarged cut with $\theta \in [0, \pi/3]$ is presented [6]

Fig. 2 Model for incompressible material behavior of trigonal symmetry in the π-plane in polar coordinates R(θ)

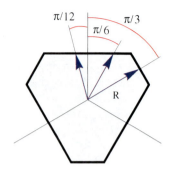

$$h = k = d = 1. \tag{9}$$

The models for incompressible behavior of trigonal symmetry are compared in the $d - k$–diagram [21–23]. For models of compressible material behavior the $\frac{1}{d} - k$ –diagram, which allows to represent the properties $d \to \infty$, $k = \sqrt{3}$ of the normal stress hypothesis, is recommended [22, 30]. The models of hexagonal symmetry for incompressible material behavior ($d = 1$) can be compared in the $h - k$-diagram (Fig. 3).

3 Yield Surfaces with Hexagonal Symmetry

The generalized models for incompressible material behavior of hexagonal symmetry in the π-plane are considered in this section:

- **Unified Yield Criterion of Yu (UYC)** [40, 44] with the parameter $b \in [0, 1]$

$$\begin{cases} \sigma_I - \dfrac{1}{1+b}(b\,\sigma_{II} + \sigma_{III}) - \sigma_{eq} = 0, \\ \sigma_I - \dfrac{1}{1+b}(b\,\sigma_{II} + \sigma_{III}) + \sigma_{eq} = 0. \end{cases} \tag{10}$$

The parameter k and h are equal to:

$$k = \sqrt{3}\,\frac{1+b}{2+b}, \qquad h = \sqrt{6}\,\frac{1+b}{1+\sqrt{3}+2b}. \tag{11}$$

The Unified Yield Criterion of Yu (Φ_{UYC}) defines the left (lower) convexity bound of the models with the hexagonal symmetry (Fig. 3). The Tresca and Schmidt-Ishlinsky criteria are obtained with $b = 0$ and $b = 1$, respectively. The model of Sokolovsky follows with $b = \dfrac{\sqrt{3}-1}{2}$.

Yield Criteria for Incompressible Materials in the Shear Stress Space 111

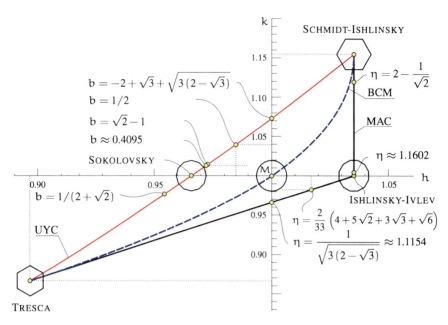

Fig. 3 h–k–diagram: models for incompressible material behavior of hexagonal symmetry: *M* model of von Mises (5) with $h = k = 1$; *UYC* Unified Yield Criterion of Yu (10); *BCM* bicubic model (12); *MAC* Multiplicative Ansatz Criterion (15). Some special points of the models are shown in order to achieve a better understanding

- **Bicubic Model** (BCM)

  This model [6, 20] is obtained as a linear combination of the Tresca (6) and Schmidt-Ishlinsky (7) criteria:

  $$(1 - \xi)\left[\left(I'_2 - \sigma^2_{eq}\right)^2 \left(2^2 I'_2 - \sigma^2_{eq}\right) - 3^3 I'^2_3\right]$$
  $$+ \xi \left[\frac{3^3}{2^3} I'_3 + \frac{3^2}{2^2} I'_2 \sigma_{eq} - \sigma^3_{eq}\right] \left[\frac{3^3}{2^3} I'_3 - \frac{3^2}{2^2} I'_2 \sigma_{eq} + \sigma^3_{eq}\right] = 0. \quad (12)$$

  The Tresca and Schmidt-Ishlinsky criteria are obtained with $\xi = 0$ and $\xi = 1$, respectively. The value $k = 1$ results in $\xi = \dfrac{2^6}{7 \times 13} \approx 0.7033$. The parameters k and h are obtained from the bicubic equations

  $$2^4 \times 3^3 + 2^3 \times 3^3 k^2 (\xi - 2^2) + 2^6 k^6 (\xi - 1) - 3^3 k^4 (7\xi - 2^4) = 0, \quad (13)$$

  $$2^5 \times 3^3 + 2 \times 3^3 h^4 (2^4 - 7\xi) + 2^4 \times 3^3 h^2 (\xi - 2^2) + h^6 (37\xi - 2^6) = 0 \quad (14)$$

  as the lowest positive solutions. The analytical solutions of (13) and (14) are complicated and hence omitted.

- **Multiplicative Ansatz Criterion** (MAC)
  The multiplicative combination of the Tresca (6) and Schmidt-Ishlinsky (7) criteria lies on the right (upper) boundary of the convexity region of the models with hexagonal symmetry (Fig. 3), see [16–18, 34]. It is obtained as follows [22]

$$\Phi_{MAC} = \left[\left(I'_2 - (\eta\,\sigma_{eq})^2\right)^2 \left(2^2 I'_2 - (\eta\,\sigma_{eq})^2\right) - 3^3 I'^2_3\right]$$
$$\times \left[\frac{3^3}{2^3}I'_3 + \frac{3^2}{2^2}I'_2\,\sigma_{eq} - \sigma^3_{eq}\right]\left[\frac{3^3}{2^3}I'_3 - \frac{3^2}{2^2}I'_2\,\sigma_{eq} + \sigma^3_{eq}\right] = 0. \quad (15)$$

The value η lies in the interval $\eta \in \left[1, \frac{4}{3}\right]$. The parameters k and h are computed to:

$$k = \frac{\sqrt{3}}{2}\eta, \quad h = \begin{cases} \eta\sqrt{3(2-\sqrt{3})}, & \eta \in \left[1, \frac{2}{\sqrt{3}}\right]; \\ \sqrt{4(2-\sqrt{3})}, & \eta \in \left[\frac{2}{\sqrt{3}}, \frac{4}{3}\right]. \end{cases} \quad (16)$$

The criteria of Tresca and Schmidt-Ishlinsky are obtained with $\eta = 1$ and $\eta = \frac{4}{3}$, respectively. With $\eta = \frac{2}{\sqrt{3}}$ the regular dodecagon in the π-plane due to Ishlinsky-Ivlev is obtained. For UYC and MAC the points, which have the shortest distance to the point M(1, 1) (Fig. 3, von Mises criterion), can be computed from the equation

$$(h-1)^2 + (k-1)^2 \to \min. \quad (17)$$

Using these points the von Mises criterion can be approximated with the dodecagons of the UYC with $b = 0.4095$ and the dodecagons of the Multiplicative Ansatz Criterion (MAC) with $\eta = 1.1344$.

- **Universal Model with Hexagonal Symmetry**
  The parameter $b \in [0, 1]$ of the UYC (10) can be replaced by the parameter $k \in \left[\frac{\sqrt{3}}{2}, \frac{2}{\sqrt{3}}\right]$ with (11)

$$b = \frac{\sqrt{3} - 2k}{k - \sqrt{3}}. \quad (18)$$

The parameter $\eta \in \left[1, \frac{4}{3}\right]$ in the MAC (15) can be replaced by $k \in \left[\frac{\sqrt{3}}{2}, \frac{2}{\sqrt{3}}\right]$ with (16)

$$\eta = \frac{2}{\sqrt{3}}k. \quad (19)$$

With the linear (convex) combination of the two latter models

$$\Phi_{12} = \xi\,\Phi_{MAC} + (1-\xi)\,\Phi_{UYC}, \quad \xi \in [0, 1] \quad (20)$$

the model with the power of the stress $n = 12$ is obtained. With two parameters (k, ξ) it covers all possible convex forms in the $h - k$ diagram. The values $k = 1$ and $\xi = 0.3901$ result in $h = 1$, which corresponds to the von Mises criterion (Fig. 3). With $\xi = 0.3901$ and $k \in \left[\frac{\sqrt{3}}{2}, \frac{2}{\sqrt{3}}\right]$ one gets the approximation of BCM (12). With $k = 1$ and $\xi \in [0, 1]$ one obtains a model, which links regular dodecagons of Sokolovsky and Ishlinsky-Ivlev: $h \in [0.9659, 1.0353]$. This model contains even powers of the equivalent stress $(\sigma_{eq})^{2i}$, $i = 0\ldots 6$ only.

4 Shear Stress Space

The models for incompressible material with hexagonal symmetry in the π-plane can be formulated in the shear stress space $(\tau_{12}, \tau_{23}, \tau_{31})$ using

$$\tau_{12} = \frac{1}{2}(\sigma_I - \sigma_{II}), \qquad \tau_{23} = \frac{1}{2}(\sigma_{II} - \sigma_{III}), \qquad \tau_{31} = \frac{1}{2}(\sigma_{III} - \sigma_I). \qquad (21)$$

The von Mises criterion (5) is represented in this space by a sphere, see [4, 26, 38] among others. The hydrostatic line is "contracted" to the point

$$\tau_{12} = \tau_{23} = \tau_{31} = 0. \qquad (22)$$

The transform has the following linear-algebraic properties: Eq. (21) allow to obtain the shear stresses from the principal stresses in the unique way. The same is not true in the case of the backward transform. It requires to consider expressions (21) as a system of linear equations with respect to the principal stresses under the assumption that the shear stresses are known. Because of the condition [26, 35]

$$\tau_{12} + \tau_{23} + \tau_{31} = 0 \qquad (23)$$

this system is not of full rank, hence it possesses an infinite number of solutions, which means that the transform is not unique. However, it is uniquely determined by the cross-section (23).

If the stresses σ_I, σ_{II} and σ_{III} are components of the stress deviator, they satisfy

$$\sigma_I + \sigma_{II} + \sigma_{III} = 0, \qquad (24)$$

and hence it follows

$$\sigma_I = \frac{2}{3}(\tau_{12} - \tau_{31}), \qquad \sigma_{II} = \frac{2}{3}(\tau_{23} - \tau_{12}), \qquad \sigma_{III} = \frac{2}{3}(\tau_{31} - \tau_{23}). \qquad (25)$$

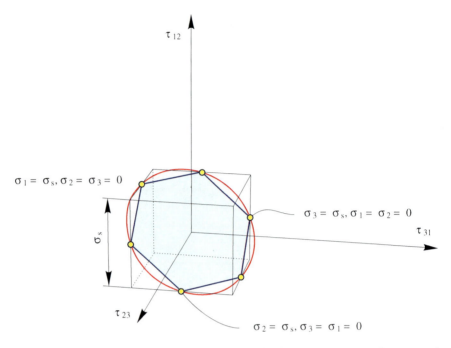

Fig. 4 Cube, the criterion of Tresca and of von Mises in the shear stress space (τ_{12}, τ_{23}, τ_{31}), after [26, 35]

For the formulation of further models Platonic[1] (regular convex polyhedron), Archimedean[2] (convex polyhedron) and Catalan[3] (Archimedean dual) solids [1, 9, 39] with orthogonal symmetry planes are used. Also superpositions of these solids can be introduced. In the cross section (23) the cut occurs, which corresponds to the shape of Φ in the π-plane. For the space filling polyhedra

- cube, isoclinal octahedron, tetrakaidecahedron (truncated octahedron) and
- rhombic dodecahedron

the criteria of Tresca and of Schmidt-Ishlinsky are obtained (Figs. 4, 5 and 6), see [44].

The triakisoctahedron [39] yields the UYC (10) with $b = \sqrt{2} - 1 \approx 0.4142$, the tetrakis hexahedron leads to $b = \dfrac{1}{2}$, the disdyakisdodecahedron—to $b = \dfrac{1}{2 + \sqrt{2}} \approx 0,2929$. The great rhombicuboctahedron [39] leads to the MAC

---

[1] http://mathworld.wolfram.com/PlatonicSolid.html
[2] http://mathworld.wolfram.com/ArchimedeanSolid.html
[3] http://mathworld.wolfram.com/CatalanSolid.html

Yield Criteria for Incompressible Materials in the Shear Stress Space

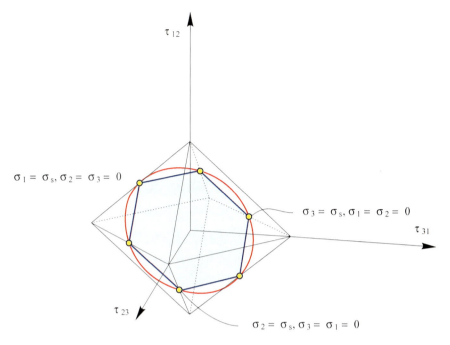

Fig. 5 Isoclinal octahedron and the criteria of Tresca and of von Mises in the shear stress space $(\tau_{12}, \tau_{23}, \tau_{31})$

(15) with $\eta = \frac{3}{14}\left(4+\sqrt{2}\right) \approx 1.1602$, the small rhombicuboctahedron yields $\eta = 2 - \frac{1}{\sqrt{2}} \approx 1.2929$, cf. [45, 46]. Further considerations of this kind can be found in [3, 19, 24, 36, 46, 48]. The models in the shear stress space are formulated in [10, 11, 13, 14, 29, 47] and generalized in [43].

5 Summary

The models for incompressible material behavior with hexagonal symmetry (4) have the property $d = 1$ (no strength differential effect). To compare these models, the relations h and k are defined as the ratios of the radii at the angles of $\theta = \frac{\pi}{12}$ and $\theta = \frac{\pi}{6}$ to the radius under the angle $\theta = 0$. They allow a uniform view of the incompressible material behavior models with hexagonal symmetry in the $h - k$-diagram.

The most widely used models are the criteria of Tresca, von Mises, and Schmidt-Ishlinsky. Together with the criteria of Sokolovsky and Ishlinsky-Ivlev

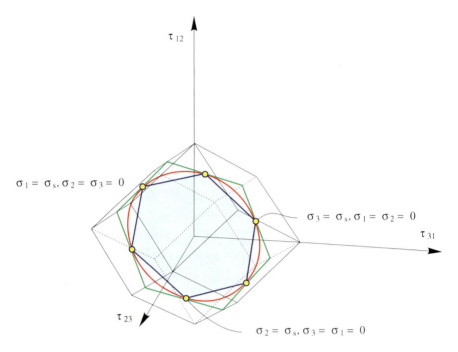

Fig. 6 Rhombic dodecahedron and the criterion of Schmidt-Ishlinsky, Tresca and of von Mises in the shear stress space (τ_{12}, τ_{23}, τ_{31})

they deliver the most important points in the $h - k$-diagram (Fig. 3). Other points of the diagram arise from the formulation of the models in the shear stress space. A physical meaning for this formulations and its connection with a material structure is not known.

The extreme lines, which connect the criteria of Tresca and Schmidt-Ishlinsky on the left and on the right of the $h - k$-diagram, arise from the UYC and the MAC. A generalized model is apparent from the convex combination of the UYC and the MAC that includes all possible convex forms in the $h - k$-diagram. This criterion has no explicit solution for σ_{eq} which leads to inconveniences upon application.

The bicubic model (BCM), which divides the $h - k$-diagram into two areas, includes the criteria of Tresca, von Mises and Schmidt-Ishlinsky. This model is continuously differentiable (excluding the borders) and allows an explicit solution for σ_{eq}. For this reason, the BCM is appropriate for practical use.

The proposed systematization simplifies the comparison and selection of criteria. This systematization shows the differences between generalized criteria (UYC, BCM and MAC) and allows the suggestion of conservative assumptions in applications.

Acknowledgments The first author was supported by the Deutsche Forschungsgemeinschaft (DFG) reference KO 3382/6-1.

A.1 6 Axiatoric-Deviatoric Invariants

The model of isotropic material behavior Φ is a function of three stress invariants, e.g. of the axiatoric-deviatoric invariants [2, 48]

$$I_1 = \sigma_I + \sigma_{II} + \sigma_{III}, \tag{26}$$

$$I'_2 = \frac{1}{2 \cdot 3}\left[(\sigma_I - \sigma_{II})^2 + (\sigma_{II} - \sigma_{III})^2 + (\sigma_{III} - \sigma_I)^2\right], \tag{27}$$

$$I'_3 = \left(\sigma_I - \frac{1}{3}I_1\right)\left(\sigma_{II} - \frac{1}{3}I_1\right)\left(\sigma_{III} - \frac{1}{3}I_1\right). \tag{28}$$

The stress angle θ [8, 28, 48], cf. [27]

$$\cos 3\theta = \frac{3\sqrt{3}}{2} \frac{I'_3}{(I'_2)^{3/2}} \tag{29}$$

is used quite often instead of I'_3.

References

1. Adam, P., Wyss, A.: Platonische und archimedische Körper, ihre sternformen und polaren gebilde. Haupt Verlag, Bern (1994)
2. Altenbach, H., Altenbach, J., Zolochevsky, A.: Erweiterte Deformationsmodelle und Versagenskriterien der Werkstoffmechanik. Deutscher Verlag für Grundstoffindustrie, Stuttgart (1995)
3. Annin, B.D.: Theory of ideal plasticity with a singular yield surface. J. Appl. Mech. Tech. Phys. **40**(2), 347–353 (1999)
4. Belyaev, N.M.: Strength of Materials. Mir Publishers, Moscow (1979)
5. Billington, E.W.: Introduction to the Mechanics and Physics of Solids. Adam Hilger Ltd, Bristol (1986)
6. Bolchoun, A., Kolupaev, V.A., Altenbach, H.: Convex and non-convex flow surfaces (in German: Konvexe und nichtkonvexe Fließflächen). Forsch Ingenieurwes **75**(2), 73–92 (2011)
7. Burzyński, W.: Studjum nad hipotezami wytężenia. Akademia Nauk Technicznych, Lwów (1928)
8. Chen, W.F., Zhang, H.: Structural Plasticity - Theory, Problems, and CAE Software. Springer, New York (1991)
9. Cromwell, P.R.: Polyhedra. Cambridge University Press, New York (1999)

10. Davis, E.A.: The Bailey flow rule and associated yield surface. J. Appl. Mech. **28**(2), 310 (1961)
11. Hershey, A.V.: The plasticity of an isotropic aggregate of anisotropic face-centered cubic crytals. J. Appl. Mech. Trans. ASME **21**(3), 241–249 (1954)
12. Hill, R.: On the inhomogeneous deformation of a plastic lamina in a compression test. Phil. Mag. Series 7 **41**(319), 733–744 (1950)
13. Hosford, W.F.: A generalized isotropic yield criterion. J. Appl. Mech. Trans. ASME **39**(June), 607–609 (1972)
14. Hosford, W.F.: On yield loci of anisotropic cubic metals. In: Proceeding of the 7th North American Metalworking Research (NAMRC), vol. 7, pp. 191–196. SME, Dearborn (1979)
15. Ishlinsky, A.Y.: Hypothesis of strength of shape change (in Russ.: Gipoteza prochnosti formoizmenenija). Uchebnye Zapiski Moskovskogo Universiteta, Mekhanika **46**, 104–114 (1940)
16. Ishlinsky, A.Y., Ivlev, D.D.: Mathematical Theory of Plasticity (in Russ.: Matematicheskaja teorija plastichnosti). Fizmatlit, Moscow (2003)
17. Ivlev, D.D.: On extremal properties of the yield criteria (in Russ.: Ob ekstremal'nych svojstvach uslovij plastichnosti). J. Appl. Math. Mech. **5**, 951–955 (1960)
18. Ivlev, D.D.: Theory of Ideal Plasticity (in Russ.: Teorija idealnoj plastichnosti). Nauka, Moscow (1966)
19. Kolosov, G.V.: On the surfaces showing the distribution of the shear stresses in a point of a continuous deformable body (in Russ.: O poverkhnostjach demonstrirujushhikh raspredelenije srezyvajushhikh usilij v tochke sploshnogo deformiruemogo tela). Prikladnya Matematika i Mekhanika **1**(1), 125–126 (1933)
20. Kolupaev, V.A.: 3D-Creep behaviour of parts made of non-reinforced thermoplastics (in German: Dreidimensionales Kriechverhalten von Bauteilen aus unverstärkten Thermoplasten). PhD thesis, Martin-Luther-Universität Halle-Wittenberg, Halle (2006)
21. Kolupaev, V.A., Altenbach, H.: Application of the generalized model of Mao-Hong Yu to plastics (in German: Anwendung der Unified Strength Theory (UST) von Mao-Hong Yu auf unverstärkte Kunststoffe. In: Grellmann, W. (ed.) Tagung Deformations- und Bruchverhalten von Kunststoffen, vol 12, pp. 320–339. Martin-Luther-Universität Halle-Wittenberg, Merseburg (2009)
22. Kolupaev, V.A., Altenbach, H.: Considerations on the Unified Strength Theory due to Mao-Hong Yu (in German: Einige Überlegungen zur Unified Strength Theory von Mao-Hong Yu). Forsch Ingenieurwes **74**(3), 135–166 (2010)
23. Kolupaev, V.A., Bolchoun, A., Altenbach, H.: New trends in application of strength hypotheses (in German: Aktuelle Trends beim Einsatz von Festigkeitshypothesen). Konstruktion, 59–66 (2009)
24. Lüpfert, H.P.: Schubspannungs-Interpretationen der Festigkeitshypothese von Huber / v. Mises / Hencky und ihr Zusammenhang. Technnische Mechanik **12**(4), 213–217 (1991)
25. Mendelson, A.: Plasticity: Theory and Application. Krieger, Malabar (1968)
26. Mises, R.V.: Mechanik des festen Körpers im plastischen deformablen Zustand. Nachrichten der Königlichen Gesellschaft der Wissenschaften Göttingen, Mathematisch-physikalische Klasse, pp. 589–592 (1913)
27. Novozhilov, V.: On the principles of the statical analysis of the experimental results for isotropic materials (in Russ.: O prinzipakh obrabotki rezultatov staticheskikh ispytanij izotropnykh materialov). Prikladnaja Matematika i Mechanika **XV**(6), 709–722 (1951)
28. Ottosen, N.S., Ristinmaa, M.: The Mechanics of Constitutive Modeling. Elsevier Science, London (2005)
29. Paul, B.: Macroscopic plastic flow and brittle fracture. In: Liebowitz, H. (ed.) Fracture: an advanced treatise, vol. II, pp. 313–496. Academic Press, New York (1968)
30. Pisarenko, G.S., Lebedev, A.A.: Deformation and Strength of Materials under Complex Stress State (in Russ.: Deformirovanie i prochnost' materialov pri slozhnom naprjazhennom sostojanii). Naukowa Dumka, Kiev (1976)
31. Prager, W., Hodge, P.: Theorie ideal plastischer Körper. Springer, Wien (1954)

32. Reuss, A.: Vereinfachte Beschreibung der plastischen Formänderungsgeschwindigkeiten bei Voraussetzung der Schubspannungsfließbedingung. Zeitschrift für Angewandte Mathematik und Mechanik **13**(5), 356–360 (1933)
33. Schmidt, R.: Über den Zusammenhang von Spannungen und Formänderungen im Verfestigungsgebiet. Ing Arch **3**(3), 215–235 (1932)
34. Shesterikov, S.A.: On the theory of ideal plastic solid (in Russ.: K postroeniju teorii ideal'no plastichnogo tela). Prikladnaja matematika i mechanika, Rossijskaja Akademija Nauk **24**(3), 412–415 (1960)
35. Sokolovsky, V.V.: Theory of Plasticity (in Russ. and English: Teorija plastichnosti). Izdatelstvo Akademii Nauk SSSR, Moscow (1946)
36. Szczepiński, W., Szlagowski, J.: Plastic Design of Complex Shape Structures. Ellis Horwood Limited and PWN-Polish Scientific Publishers, Chichester, Warszawa (1990)
37. Tresca, H.: Mémoire sur l'ecoulement des corps solides. Mémoires pres par div sav **18**, 75–135 (1868)
38. Walczak, J.: Nowoczesna miara wytężenia materiału. Archiwum mechaniki stosowanej **III**, 5–26 (1951)
39. Wolfram, S.: The Mathematica Book: the Definitive Best-Selling Presentation of Mathematica by the Creator of the System. Wolfram Media, Champaign (2003)
40. Yu, M.H.: General behaviour of isotropic yield function (in Chinese). Scientific and technological research paper of Xi'an Jiaotong University, pp. 1–11 (1961)
41. Yu, M.H.: Twin shear stress yield criterion. Int J Mech Sci **25**(1), 71–74 (1983a)
42. Yu, M.H.: Twin shear stress yield criterion. Int J Mech Sci **25**(11), 845–846 (1983b)
43. Yu, M.H.: Engineering Strength Theory (in Chinese). Higher Education Press, Beijing (1999)
44. Yu, M.H.: Unified Strength Theory and its Applications. Springer, Berlin (2004)
45. Yu, M.H.: Linear and non-linear unified strength theory (in Chinese). J Geotech Eng **26**(4), 662–669 (2007)
46. Yu, M.H., Xia, G., Kolupaev, V.A.: Basic characteristics and development of yield criteria for geomaterials. J Rock Mech and Geotech Eng **1**(1), 71–88 (2009)
47. Zhou, X.P., Zhang, Y.X., Wang, L.H.: A new nonlinear yieid criterion. J of Shanghai Jiaotong University (Science) **E-9**(1), 31–33 (2004)
48. Życzkowski, M.: Combined Loadings in the Theory of Plasticity. PWN-Polish Scientific Publishers, Warszawa (1981)

The Optimum Design of Laminated Slender Beams with Complex Curvature Using a Genetic Algorithm

Jun Hwan Jang and Jae Hoon Kim

Abstract The chapter presents the optimum structural design for composite slender beams with complex curvature. The optimization process is performed using a genetic algorithm (GA), associated with a variational asymptotic method for the structural analysis. The stiffness control of arbitrary, complex sections under some design conditions is performed for composite beam where the geometrically nonlinear characteristic of the structure is considered. The objective function was defined as the weight, strength and fatigue life. The laminate thicknesses are to be determined optimally by defining a design index comprising a weighted average of the objective functions and determining the minimum.

Keywords Slender beam · Genetic algorithm · Coupled stiffness · Variational asymptotic method

1 Introduction

Among the rotor system of helicopter, the rotor blade is an essential component that generates lift, thrust, and control force required for flight. Since it has to endure the centrifugal force and lift produced by rotation and also has to release the lift unbalance and blade weight via flapping, feathering, and lead-lag movement, certain features (lightweight, high strength, expediency in manufacture via all-in-all mold, and performance enhancement) are required. Blade manufacture is

---

J. H. Jang · J. H. Kim (✉)
Chungnam National University, 99 Daehak-ro, Yuseong-gu, Daejeon, South Korea
e-mail: kimjhoon@cnu.ac.kr

J. H. Jang
e-mail: bulbearj@empal.com

thus one of the fields that utilize various composite materials and corresponding high-tech manufacture procedures most frequently.

A composite rotor blade generally consists of many structures—such as skin, spar, honeycomb, form that sustains airfoil shape, weight balance, de-icing device, and erosion shield at the leading edge of airfoil. Since a rotor blade requires many special features such as high strength with lightweight, manufacture using consolidated mold, and performance enhancement, various composite materials are used. Moreover, a rotor blade also needs a warranty for fatigue life under differing loads following numerous operation conditions. High frequency cyclic stress inflicted on a rotor blade, which originates from the cyclic aerodynamic load that follows rotation, can result in serious failure. Therefore, as well as static stability, fatigue characteristics and vibration characteristics of a blade must be taken into account.

The analyses methods of rotor blade have undergone considerable progress through long time. These progresses were mainly accomplished by the analysis with the finite element method. The nonlinear modeling method that could explain couple and transverse shear deformation [1, 2], was the first to be developed to describe the structural static behavior of rotor blades made up of composite materials. Before the development of the nonlinear modeling method, it was difficult to apply the 3-D material coefficient to section modulus equation since most of classic theories could not deal with phenomena such as elastic coupling in anisotropic material and initial twist or curvature. From the 1980s on, significant progresses have been made on the nonlinear modeling method. Giavotto, Borri and Mantegazza [3] pioneered section analysis of ordinary composite beam using framework based on linear elastic theory. Although transverse shear effect was not properly reflected, Kosmatka [4] and Friedmann [5] dealt with geometrical nonlinear problems for the first time and also applied it to dynamic nonlinear problems of the rotor blade.

The variational asymptotic method (VAM) of Berdichevsky [6] is a very useful mathematical tool that simplifies the process of finding static points that depend on one or more small variables. A function that finds static points from proper variables can be applied to any other point-finding problems. Moreover, it makes it possible to divide the analysis of a 3-D nonlinear structure such as a beam into nonlinear 1-D analysis or linear 2-D analysis. A 6 × 6 and 4 × 4 sectional stiffness and mass matrix can be derived this way. It can also be applied to thin wall beams composed of composite materials as well as simple isotropic beam. Hodges [7] developed the variational asymptotic beam sectional analysis of the Euler beam based on the VAM theory, and Wenbin [8] applied it the Timoshenko model.

The variational asymptotic beam sectional analysis (VABS) [9, 10] can simplify a 3-D model with large aspect ratio, such as a blade, to enable modeling of coupled characteristics, and is also able to calculate with high accuracy in a short time. Among many theories about arbitrary beam analysis, VABS is the most accurate, and is able to acquire various kinds of cross-sectional information. It is a program optimized for rotor blade analysis/design, capable of 3-D displacement, strain, and stress analysis using cross-sectional information. It utilizes the calculated stiffness matrix and mass matrix, and many previous studies have shown that results obtained from this kind of method matches that obtained from dividing standard nonlinear 3-D beam problem into 2-D or 1-D [11].

More than one criterion is needed to design a composite blade. That is, an optimization process is needed to satisfy all the criteria expressed as a multi-objective function. Multi-objective optimization is an optimization method that can solve problems in which several criteria cannot be combined into one criterion. Genetic algorithm is effective for obtaining the solution of multi-objective optimization, because it does not need a gradient and it obtains several solutions rather than one local solution.

2 Genetic Algorithm-Applied Study

Genetic algorithms, due to their simplicity and generality, are utilized in many fields. Starting from the design of mass spring dashpot systems done by Goldberg [12] using a simple genetic algorithm (Simple-GA), numerous studies are in progress until today. Studies using GA have a broad range of applications, including optimization of pipelines, behavioral evolution of robots, learning process of neural circuits, and optimization of fuzzy membership functions. B. P. Wang [13] introduced the genetic algorithm to minimize the spring-in phenomenon resulting from residual stress that is induced when a reinforced composite material cools down. Hajela [14] showed characteristics of the genetic algorithm, in the case of using it for the design of multidisciplinary rotor blades with non-convex design variables. Rodolphe Le Riche [15] optimized the laminate stacking sequence for improved buckling load. In this optimization, permutation, a new genetic operator, was suggested and proven to be efficient. A permutation operator was also used by Boyang Liu [16] for optimization of stacking sequence. Here, using a genetic algorithm, we developed an algorithm that is capable of establishing a design that realizes weight reduction of blades while also satisfying several criteria.

3 Genetic Algorithm

The genetic algorithm is based on genetics and principles of biological evolution. It is an algorithm that puts into practice the rule of nature, that individuals with superior characteristics adapt better to natural environment and produce superior progeny. It is therefore very different from classical approaches. Classical optimization methods used approaches such as searching objective functions of every point in a limited space one by one utilizing the gradient, or choosing an arbitrary point and starting to search from there. Thus, classical optimization was efficient in a narrow domain, but was not in a broad domain.

However, since genetic algorithms do not utilize the concept of gradient at all and performs directional search and probability search, they have an advantage that they can solve optimization problem comprising features like combination of continuity-discreteness, discontinuity, and non-convex domain. The genetic algorithm optimizes a population of individuals composed of binary digits, via three processes—selection, crossover, and mutation. Selection, crossover, and mutation are the most fundamental operators in a genetic algorithm. In Fig. 1 it shows the structure of the genetic algorithm.

Selection is a process where individual strings are replicated in accordance with its fitness. Thus, it plays the role of delivering characteristics of strings with high fitness to the next generation in the course of evolution. During crossover, parts of strings of selected chromosomes, which are equivalent to information, are exchanged. Crossover gives chances to obtain high fitness to chromosomes and accelerates converge speed. Lastly, mutation is a process where information not available through selection and crossover can be obtained. With proper probability, the mutation process results in searching effect in the whole area, thus increases convergence to high fitness to a great extent. Operators like crossover and mutation play the role of increasing diversity of a population.

4 Laminated Slender Beam Geometry for Structural Optimization

Development tendencies of rotor blades can be grouped into 3 generations. The 1st generation consists of a metal blade with rectangular shaped tip, while the 2nd generation consists of composite blade with rectangular tip. The 3rd generation blade, also made up of composite materials, has a revised tip shape in order to

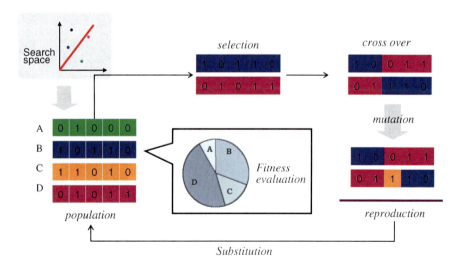

Fig. 1 Structure of genetic algorithm

obtain performance enhancement along with reduction of noise, vibration, and take-off weight. Here, we applied our study of the 3rd generation blade, a blade with excellent characteristics such as maximized aerodynamic performance and reduced noise and vibration. Figure 2 illustrates the shape of the laminated slender beam and the location of airfoil. The tail rotor that is the subject of the optimal design has total length of 1575 mm (measured from the center of hub) and is composed of two airfoils.

5 Load Condition

The blade strength and stiffness should be between $-0.5 \sim +3.5$ g, the structural load coefficients of "Class I" helicopter stated in MIL-S-8698 [17]. We set up the centrifugal force as 300 kN. Since the flap-wise moment exerts the greatest influence on the airfoil design, this value should be decided with careful consideration. Since the torsion of the blade and the stiffness of lag-wise are relatively large, the lead-lag moment and torsion moment do not have significant influence of the static structure design, but they can have great importance in the vibratory and aeroelastic aspect. Within flight condition and several multiples of design limit speed, the flutter and divergence phenomena should not happen to the rotor blade, and no aeroelastic instability is acceptable within several multiples of limit speed of the rotor blade design, in both power-on and autorotation state. The load of flight condition was generated using CAMRAD II [18], commercial software for aerodynamical analysis of rotorcraft, and the limit load of design is based on ADS-29. In Figs. 3, 4, 5 several load conditions for the generation of the preliminary design load of the tail rotor component were also considered. Case of torque-up is the isolated rotor model, full

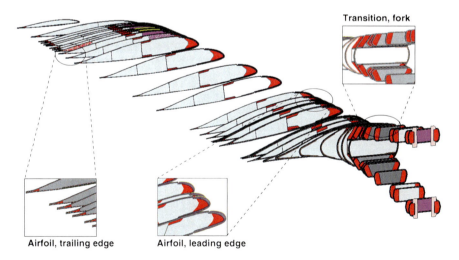

Fig. 2 Laminated slender beam configuration for structure optimization

thrust as defined by full control of +17.5° blade pitch. And the case of speed-up is the isolated rotor model, 1521 rpm, blade pitch angle of −18.5°, associated thrust and power.

6 Optimized Design for Laminated Slender Beam

6.1 Nonlinear Slender Beam Modeling for Optimized Design of Blade Structures

Generally, in structural design of airplanes, finite element analysis is used for accurate stress and vibration analysis. However, building a 3-D finite analysis model—that is made almost analogous to the real structure to enable delicate analysis—consumes a lot of time and human resources, and much knowledge and know-how is also needed to build reliable analysis models. In addition, in case of delicate 3-D models, extraction of data needed for applied analysis such as understanding of qualitative physical tendency or aeroelasticity analysis often yields unsatisfactory results. Moreover, as in preliminary design stages, when definite data about structural and aerodynamic shape is not sufficiently provided and shape change is frequent, modifying the 3-D structural model accordingly to every change is almost impossible. A composite blade has domains that undergo radical changes in geometrical shape and also domains that undergo changes in material properties. Building a 3-D model in preliminary design is thus practically unrealistic. For efficient static, dynamic, and fatigue analysis, equivalent modeling, which can describe the macroscopic behavior relatively easily but efficiently, is required [19–22]. Here, we perform an equivalent modeling in the preliminary design stage, using variational asymptotic beam sectional analysis to assume the elastic axis and its value of EI, GJ, coupled stiffness and mass. In Fig. 6 equivalent modeling is performed by dividing the blade into several domains after selecting parts which undergoes change in shape and which undergoes addition or reduction of material,

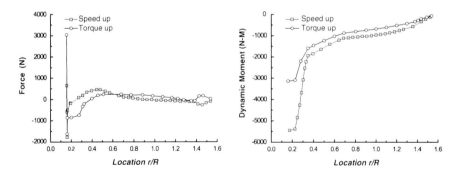

Fig. 3 Lift moment and drag force

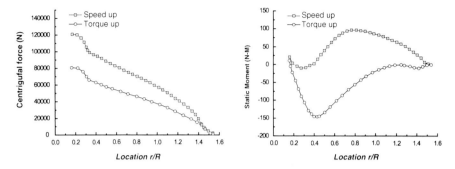

Fig. 4 Centrifugal force and flap moment

Fig. 5 Lead-lag moment

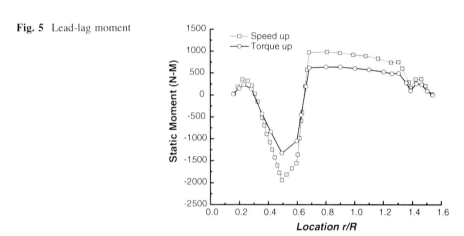

and the calculation of the coupled stiffness is performed using the calculation done on each domain for the equivalent lumped mass and mass inertia moment.

6.2 Variational Asymptotic Beam Sectional Analysis

A rotor blade has its length much greater than the other two dimensions and thus has often been treated as a beam, a 1-D structure, to reduce the computational costs associated with the analysis [7, 23, 24]. In order to perform this ideal situation without loss of accuracy, one has to capture the behavior associated with the two omitted dimensions (the cross sectional coordinates) by correctly accounting for geometry and material distribution. VABS is able to calculate the one-dimensional cross-sectional stiffness constants, with transverse shear and Vlasov refinements, for initially twisted and curved beams with arbitrary geometry and material properties. The variational asymptotic method (VAM) developed by Berdichevsky

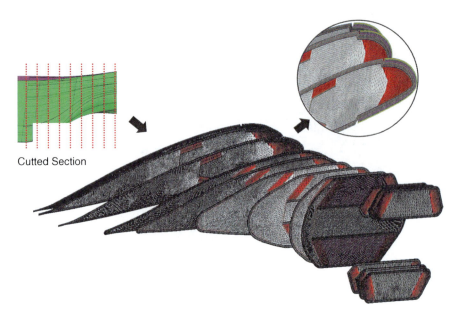

Fig. 6 Nonlinear beam's sectional modeling with curvature

is the mathematical basis of VABS and is used to split a general 3-D nonlinear elasticity problem for a beam-like structure into a two-dimensional (2-D) linear cross-sectional analysis and a 1-D nonlinear beam analysis.

VABS allows one to replace a 3-D structural model with a reduced-order model in terms of an asymptotic series of certain small parameters inherent to the structure. These small parameters are h/l and h/R, where h is the characteristic length of the cross-section, l is the characteristic wavelength of deformation along the longitudinal direction, and R is the characteristic radius of the initial curvature and twist. The main small parameter for straight blades is h/l. The generalized strain energy per unit length accounting for transverse shear and trapeze effect is shown below [25];

$$U = \frac{1}{2} \begin{Bmatrix} \gamma_{11} \\ 2\gamma_{12} \\ 2\gamma_{13} \\ \kappa_1 \\ \kappa_2 \\ \kappa_3 \end{Bmatrix}^T \begin{bmatrix} S_{11} & S_{12} & S_{13} & S_{14} & S_{15} & S_{16} \\ S_{12} & S_{22} & S_{23} & S_{11} & S_{25} & S_{26} \\ S_{13} & S_{23} & S_{33} & S_{11} & S_{35} & S_{36} \\ S_{14} & S_{24} & S_{34} & S_{44} & S_{45} & S_{46} \\ S_{15} & S_{25} & S_{35} & S_{45} & S_{55} & S_{56} \\ S_{16} & S_{26} & S_{36} & S_{46} & S_{56} & S_{66} \end{bmatrix} \begin{Bmatrix} \gamma_{11} \\ 2\gamma_{12} \\ 2\gamma_{13} \\ \kappa_1 \\ \kappa_2 \\ \kappa_3 \end{Bmatrix}$$

$$+ \begin{Bmatrix} \gamma_{11} \\ \kappa_1 \\ \kappa_2 \\ \kappa_3 \end{Bmatrix}^T (\gamma_{11}A + \kappa_1 B + \kappa_2 C + \kappa_3 D) \begin{Bmatrix} \gamma_{11} \\ \kappa_1 \\ \kappa_2 \\ \kappa_3 \end{Bmatrix} \qquad (1)$$

6.3 Nonlinear, Intrinsic Beam Equations

The nonlinear, intrinsic, mixed equations for the dynamics of a general (non-uniform, twisted, curved, anisotropic) beam undergoing small strains and large deformation are given below [26, 27].

$$F' + (\tilde{k} + \tilde{\kappa})F + f = \dot{P} + \tilde{\Omega}P$$

$$M' + (\tilde{k} + \tilde{\kappa})M + (\tilde{e}_1 + \tilde{\gamma})F + m = \dot{H} + \tilde{\Omega}H + \tilde{V}P$$

$$V' + (\tilde{k} + \tilde{\kappa})V + (\tilde{e}_1 + \tilde{\gamma})\Omega = \dot{\gamma}$$

$$\Omega' + (\tilde{k} + \tilde{\kappa})\Omega = \dot{\kappa} \qquad (2)$$

where $(\ )'$ denotes the partial derivative with respect to the axial coordinate of the undeformed beam, and $\dot{(\ )}$ denotes the partial derivative with respect to time. $x(x,t)$ and $M(x,t)$ are the measure numbers of the internal force and moment vector (cross-section stress resultants), $P(x,t)$ and $H(x,t)$ are the measure numbers of the linear and angular momentum vector (generalized momenta), $\gamma(x,t)$ and $\kappa(x,t)$ are the beam strains and curvatures (generalized strains), $V(x,t)$ and $\Omega(x,t)$ are the linear and angular velocity measures (generalized speeds), and $f(x,t)$ and $m(x,t)$ are the external force and moment measures. Measure numbers of all variables except for k are calculated in the B frame, i.e. the deformed beam cross-sectional frame. $k(x)$ is the initial twist/curvature of the beam. The measure numbers of k are in the undeformed beam cross-sectional frame.

The cross-sectional stress resultants are related to the generalized strains via the cross-sectional beam stiffnesses/flexibilities. These cross-sectional properties can be calculated using an analytical thin-walled theory [28] or computational FEM analysis [10] for general configurations. Such an analysis gives the following linear constitutive law,

$$\begin{Bmatrix} \gamma \\ \kappa \end{Bmatrix} = \begin{bmatrix} R & S \\ S^T & T \end{bmatrix} \begin{Bmatrix} F \\ M \end{Bmatrix} \qquad (3)$$

where $R(x)$, $S(x)$, and $T(x)$, are the cross-sectional flexibilities of the beam. This linear constitutive law is valid only for small strain, but the global deformations still may be large. The generalized momenta are related to the generalized speeds via the cross-sectional beam inertia,

$$\begin{Bmatrix} P \\ H \end{Bmatrix} = \begin{bmatrix} \mu\Delta & -\mu\tilde{\xi} \\ \mu\tilde{\xi} & I \end{bmatrix} \begin{Bmatrix} V \\ \Omega \end{Bmatrix} = \begin{bmatrix} G & K \\ K^T & I \end{bmatrix} \qquad (4)$$

where $\mu(x)$, $\xi(x)$, $I(x)$ are the mass per unit length, mass center offset (vector in the cross-section from the beam reference axis to the cross-sectional mass center), and mass moment of inertia per unit length, respectively. Usually, the constitutive laws are used to replace some variables in terms of others. Here it was decided to express the generalized strains in terms of the cross-section stress resultants, allowing easy specification of zero flexibility, and the generalized momenta in terms of generalized speeds, allowing easy specification of zero inertia. Thus, the primary variables of interest are F, M, V and Ω.

7 Optimization Strategy for Composite Blade

7.1 Local Optimization

The structure of the composite blade can be divided into three geometrically distinct parts—upper and lower arm that carry centrifugal force, transition domain where the shape changes radically, and airfoil. Table 1 shows some information of shape of laminated slender beam. Figure 7 illustrates the shape of a section through the composite blade. While developing the optimal design of the blade section, we supposed that a composite blade consists of 2 torsion boxes, form, and skin. Placed anterior to the torsion box is a glass UD that carries main load of blade. In the case of skin, although the ply sequence of the composite materials can be optimized in accordance with load. A common ply sequence is applied instead since it is relatively thinner and more conveniently manufactured. Design variables were defined to be the skin thickness, width of UD glass, thickness of stiffener, and location of form. Figure 6 illustrates the use of the finite element model for the calculation of the coupled stiffness matrix and coupled mass matrix using the variational asymptotic beam sectional analysis.

The centre of mass is slightly off the generalized shear center and is located at $x_2 = -18.83$ mm and $x_3 = 3.11$ mm. The generalized Timoshenko model obtained from VABS for this blade section is represented by the following matrix of the cross-sectional stiffness constants:

$$S_{r/R=0.443} = \begin{bmatrix} 6.121E+07 & -8.206E-11 & -1.756E-11 & -1.898E-09 & -6.193E+05 & 2.634E+07 \\ -8.206E-11 & 7.148E+06 & -2.269E+05 & -1.723E+07 & 1.066E-09 & 3.154E-09 \\ -1.756E-11 & -2.269E+05 & 9.111E+05 & -7.353E+06 & 1.374E-10 & 9.393E-10 \\ -1.898E-09 & -1.723E+07 & -7.353E+06 & 4.370E+09 & 1.543E-08 & 8.914E-08 \\ -6.193E+05 & 1.066E-09 & 1.374E-10 & 1.543E-08 & 6.323E+09 & -1.218E+09 \\ 2.634E+07 & 3.154E-09 & 9.393E-10 & 8.914E-08 & -1.218E+09 & 2.014E+11 \end{bmatrix}$$

(5)

And we calculate the flexibility matrix by calculating the inverse of the Timoshenko's stiffness matrix.

$$F_{r/R=0.443} = \begin{bmatrix} 1.633E-08 & 2.235E-25 & 4.433E-25 & 8.762E-27 & 1.189E-12 & -2.129E-12 \\ 2.235E-25 & 1.427E-07 & 4.063E-08 & 6.310E-10 & -2.703E-26 & -2.896E-27 \\ 4.433E-25 & 4.063E-08 & 1.124E-06 & 2.051E-09 & -3.762E-26 & -7.071E-27 \\ 8.762E-27 & 6.310E-10 & 2.051E-09 & 2.347E-10 & -7.478E-28 & -1.289E-28 \\ 1.189E-12 & -2.703E-26 & -3.762E-26 & -7.478E-28 & 1.5833-10 & 9.576E-13 \\ -2.129E-12 & -2.896E-27 & -7.071E-27 & -1.289E-28 & 9.5761-13 & 4.969E-12 \end{bmatrix}$$

(6)

Table 1 Material Information of Laminated Slender Beam Structure

Material	E_1, E_2, G_{12}(MPa)	v_{12}, v_{12}	Thickness (mm)
Carbon-Epoxy Fabric	54000, 37800, 3730	0.3, 0.3	0.36
			0.36
Carbon-Epoxy Unidirectional Tape	131000, 19650, 4800	0.42, 0.42	0.13
Glass–Epoxy Fabric	19600, 13720, 3040	0.3, 0.3	0.31
			0.31
Glass–Epoxy Unidirectional Tape	54690, 8190, 5870	0.31, 0.31	0.25
Titanium	190000, 190000, 73400	0.3, 0.3	0.34

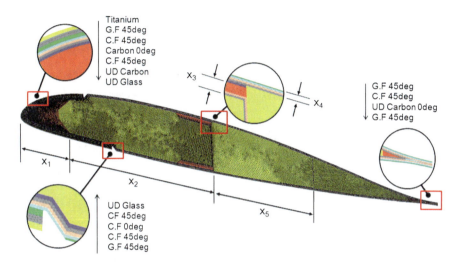

Fig. 7 Design variables and ply sequence configuration

7.2 Global Optimization

Design variables needed at preliminary stage of the blade design can be determined by values of couple and stiffness obtained after performing the equivalent modeling. The number of finite element model elements and nodes should be sufficiently large in order to perform nonlinear equivalent modeling. A smaller number of elements and panel points are critical for saving time for the calculation, but enough elements are needed for accurate determination of coupled characteristics in the equivalent modeling. Each coupled stiffness matrix and mass matrix was placed in span direction of the blade and was analyzed for global vibration. A modal analysis in vacuum was done for identification of frequency. Figure 8 shows the result of the global optimization for the normal mode.

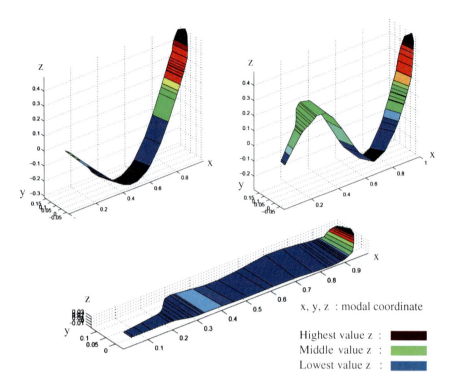

Fig. 8 Global optimization, normal mode results

The objective of the modal analysis was to verify if the separation of frequency in the fan plot between 3/rev, 4/rev, and 5/rev is enough. To ensure dynamic stability, the coupled stiffness matrix was controlled by modification of the number of design variables. Finally, the optimization of the design variables set at the beginning was done for designing a more lightweight composite blade.

8 Optimization Procedure

We utilized the genetic algorithm for the analysis of the composite blade structure optimization. When optimizing a design made up of composite materials, a genetic algorithm has an advantage in such a way that it can be easily applied to both continuous variables and discrete variables such as ply number and ply angle [29–31].

We selected the mass of the blade as an ultimate objective of optimization of the laminated slender beam section. We selected the thickness and location of the skin and torsion box as design variables and as constraints the static strength, mass center, fatigue life, and natural vibration mode.

Figure 9 shows the flow chart of the optimization method used. When the initial design value is determined, the cross-sectional analysis is performed in accordance

The Optimum Design of Laminated Slender Beams

with the location of the spanwise on the composite blade. To do this, a pre-process for finite element analysis should be done on every section. Next, using a program, the equivalent properties of the blade is calculated. Stiffness matrix and mass matrix are included in equivalent properties of each section that had been set at the beginning. The genetic algorithm searches by permuting disposition information and material properties from initial design variables. Regarding local optimization, the required strength of composite materials and required fatigue life should be satisfied by the load applied on each section. Regarding global optimization, the blade's vibration condition should be satisfied and stiffness should be controlled in order to accomplish weight reduction. To achieve a converged optimization result, the genetic algorithm should be repeated in finite loop to make it converge. We repeated 250 loops for the evaluation of convergence and then extracted the stiffness from the diagonal elements of coupled stiffness matrix for the same purpose.

Objective Function: Find $\{X\} = \{X_1, X_2, X_3, X_4 \ldots \ldots X_N\}$
Objective: Blade weight, Avoid divergence
Bound of Design Variables
Forward Glass–Epoxy UD Area: $50 < X_1 < 90$
Forward torsion box range: $60 < X_2 < 140$
Backward Glass–Epoxy UD Thickness: $1.0 < X_3 < 3.0$
Skin thickness: $0.25 < X_4 < 4.0$
Backward torsion box thickness: $55 < X_5 < 85$
Torsion box position : $120 < X_6 < 180$
Constraints
Maximum stress failure criteria

$$\sigma_1^C < \sigma_1 < \sigma_1^T$$
$$\sigma_2^C < \sigma_2 < \sigma_2^T \qquad (7)$$
$$|\tau_{12}| < \tau_{12}^S$$

Maximum strain failure criteria

$$\varepsilon_1^C < \varepsilon_1 < \varepsilon_1^T$$
$$\varepsilon_2^C < \varepsilon_2 < \varepsilon_2^T \qquad (8)$$

Fatigue life > 10000 flight hour.

$$\frac{S}{S_{endurance\ limit}} = \frac{A}{N^\gamma}, D_i = n_i/N_i,$$

$$\text{Required life} > 10000 \qquad (9)$$

Failure index of each material < 1.0
Mass center
$25\ \%\text{Chord} - 10\ \% <$ Mass center $25\ \%\text{Chord} + 10\ \%$.
Natural frequency: frequency separation at 3/rev, 4/rev, 5/rev.

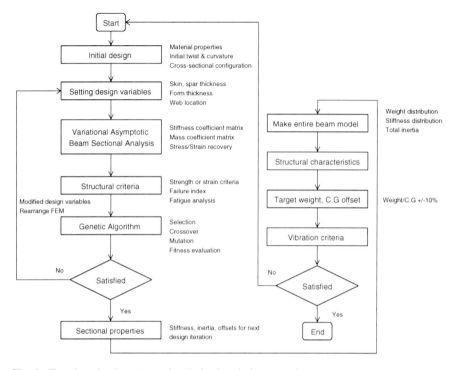

Fig. 9 Flowchart for the structural optimization design procedure

$$([K] - \omega^2[M])\{\delta\} = 0 \qquad (10)$$

9 Result of Optimized Design Derived from Genetic Algorithm

The genetic algorithm does not search individual points but searches population of design points in several discrete design spaces at once. In other words, it is able to do a local search in one space while searching in many other spaces at the same time. While methods depending on gradient at individual points have a drawback that they sometimes fall into local optimal points, the genetic algorithm is not sensitive to problems of this kind that occur in the complex design space, since it searches the population of design points, not individual points. The genetic algorithm yields not a single solution for the problem but a group of solutions which can differ from each other. The shape of the initial section of the laminated slender beam section for optimization is as shown in Fig. 10. Optimized design variables of the blade section are shown in Table 2.

Fig. 10 GA optimization results, fatigue life at r/R = 0.654

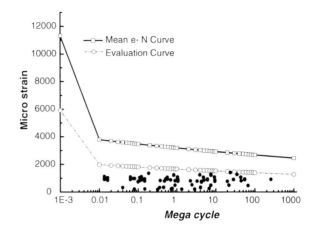

Table 2 Optimization Results of Composite Blade using Genetic Algorithm

r/R	X_1 Initial	X_1 Final	X_2 Initial	X_2 Final	X_3 Initial	X_3 Final	X_4 Initial	X_4 Final	X_5 Initial	X_5 Final
0.274	35.0	31.2	71.9	75.7	5.0	4.6	3.0	2.5	75.0	70.8
0.291	35.0	29.3	71.9	77.6	5.0	4.0	3.0	2.6	75.0	69.4
0.295	35.0	28.9	72.3	78.4	5.0	3.9	3.0	2.4	75.0	68.7
0.307	35.0	29.0	74.1	80.1	5.0	3.9	3.0	2.2	75.0	67.5
0.361	35.0	23.1	78.1	90.0	5.0	2.9	3.0	2.6	75.0	59.1
0.381	35.0	25.4	78.9	88.5	5.0	3.2	3.0	2.2	75.0	61.7
0.386	35.0	25.3	79.2	88.9	5.0	2.2	3.0	2.1	75.0	61.5
0.401	35.0	21.9	77.9	91.0	5.0	2.3	3.0	1.9	75.0	60.8
0.421	35.0	21.7	80.2	93.5	5.0	1.7	3.0	1.4	75.0	60.0
0.452	35.0	21.7	79.9	93.2	5.0	1.7	3.0	1.2	75.0	60.0
0.557	35.0	21.7	79.9	93.2	5.0	1.7	3.0	1.2	75.0	60.0
0.634	35.0	21.7	76.4	89.7	5.0	1.7	3.0	1.1	75.0	63.5
0.738	35.0	21.4	79.4	93.0	5.0	1.7	3.0	1.2	75.0	60.0
0.846	35.0	29.9	78.4	83.5	5.0	1.8	3.0	1.0	75.0	61.0
0.879	35.0	31.8	78.1	81.3	5.0	1.8	3.0	1.1	75.0	60.0
0.889	35.0	31.4	78.0	81.6	5.0	2.0	3.0	1.1	75.0	60.0
0.897	35.0	36.9	79.4	77.5	5.0	1.0	3.0	1.1	75.0	60.0
0.916	35.0	29.9	77.9	83.0	5.0	1.0	3.0	1.1	75.0	61.5
0.933	35.0	29.7	73.0	78.3	5.0	1.0	3.0	1.4	75.0	64.9

The constraint of optimization has a failure index less than 1.0, and the center of mass is located at 25 % of chord. In Figs. 3, 4 and 5, the maximum load is a force of 120 kN in the direction of blade axis and the moment of Max. 1000 N-M, Min. −2000 N-M in the direction of leadlag wise. The result of the blade section shape optimization using the genetic algorithm is shown in Table 2. Through optimization, the initial mass of the blade could be reduced from 7.35 kg to

Fig. 11 GA optimization results, modal analysis in vacuum

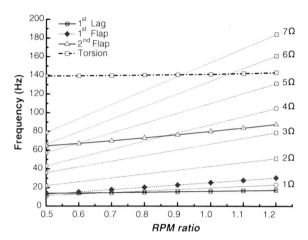

6.55 kg. The center of mass is located at 625.8 mm from the laminated slender beam's reference. Figure 11 compares the shape of the laminated slender beam section before and after optimization. The life evaluation of the rotorcraft is calculated using the safe-life method or fail-safe method. The fatigue life calculated applying the load spectrum came out to be infinite in all domains except for $r/R = 0.07$ and $r/R = 0.654$, even though it was initially set to satisfy 40000 h in each section. After optimization, the skin became relatively thinner and front/rear torsion box became both smaller and thinner. The blade mass changed for 10.9 % of the amount, from 7.35 kg to 6.55 kg. The front torsion box became bigger while the rear torsion box became smaller. Distribution of mass, before and after optimization, is illustrated in Fig. 12, respectively.

Fig. 12 GA optimization results, weight reduction accord to location

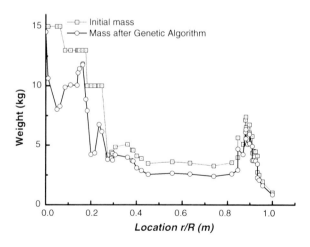

10 Conclusion

We described procedures of optimization of composite blade design that utilizes a genetic algorithm, decomposed into a local and global domain. In the optimization process, we controlled the stiffness to locally satisfy structural requirements of composite materials and globally satisfy vibration requirements. The stacking sequence of the skin of given geometry was not considered, since the elements that support the main load are UD glass, UD carbon, and carbon fabric. Thus, the thickness of the thin outer skin is not necessarily a main element of an optimal design.

1. Asymptotic beam sectional analysis was used for calculation of section coefficients of the coupled stiffness matrix. In the case of a blade that undergoes changes in shape and internal material properties, an equivalent beam model should be used. Later we calculated the static structural safety by restoring the 1-D model to a 2-D model, done through stress recovery. As an objective function of optimization, the strain-cycle curves of each composite material that constitute the blade were used.
2. For the optimization of each cross-section through the genetic algorithm, the objective function, design variables, and constraints were fixed in order to have failure indices of all objective functions less than 1.0. The cross-section shape optimization accomplished approximately a 10.9 % weight reduction of the blade.
3. Cross section shapes of local area optimized for each respective section were subject to an optimization in global area, considering the vibration condition. The stiffness matrix was optimized in order to prevent flutter and divergence phenomena from occurring on the composite blade in flight course conditions and several multiples of limit speed of design.
4. Weight reduction, the ultimate goal of airplane structure optimization, was accomplished. After optimization, the initial mass of the blade was reduced from 7.35 kg to 6.55 kg.

References

1. Kärger, L., Wetzel, A., Rolfes, R., Rohwer, K.: A three-layered sandwich element with improved transverse shear stiffness and stresses based on FSDT. Comput. Struct. **84**, 843–854 (2006)
2. Sathyamoorthy, M.: Effects of large amplitude, transverse shear and rotatory inertia on vibration of orthotropic elliptical plates. Int. J. Non-Linear Mech. **16**, 327–335 (1981)
3. Giavotto, V., Borri, M., Mantegazza, P., Ghiringhelli, G., Carmaschi, V., Maffioli, G., Mussi, F.: Anisotropic beam theory and applications. Comput. Struct. **16**, 403–413 (1983)
4. Kosmatka, J., Dong, S.B.: Saint-venant solutions for prismatic anisotropic beams. Int. J. Solids Struct. **28**, 917–938 (1991)

5. Friedman, Z., Kosmatka, J.B.: Exact stiffness matrix of a nonuniform beam II. Bending of a timoshenko beam. Comput. Struct. **49**, 545–555 (1993)
6. Berdichevskii, V.L.: Variational-asymptotic method of constructing a theory of shells. Appl. Math. Mech. **43**(4), 664–687 (1979)
7. Hodges, D.H.: Nonlinear composite beam theory. AIAA **41**, 1131–1137 (2006)
8. Yu, W., Hodges, D.H., Volovoi, V., Cesnik, C.E.S.: On Timoshenko-like modeling of initially curved and twisted composite beams. Int. J. Solids Struct. **39**, 5101–5121 (2002)
9. Yu, W.: VABS manual for users (2010)
10. Cesnik, C.E.S., Hodges, D.H.: VABS: a new concept for composite rotor blade cross-sectional modeling. JAHS **42**, 27–38 (1997)
11. Yeo, H., Ormiston. R.A.: Assessment of 1-d versus 3-d methods for modeling rotor blade structural dynamics, 51st AIAA Conference (2010)
12. Goldberg, D.E.: Genetic algorithms in search, operation, and machine learning. Addison-Wesley Publishing Company (1989)
13. Wang, B.P., Tho, C.H., Henson, M.: Minimization of spring-in composite angle components by genetic algorithm. AIAA **97**, 2733–2738 (1997)
14. Hajela, P., Lee, J.: Genetic algorithms in multidisciplinary rotor blade design. AIAA **95**, 2187–2197 (1995)
15. Riche, R.L., Haftka, R.T.: Optimization of laminate stacking sequence for buckling load maximation by genetic algorithm. AIAA **31**, 951–956 (1993)
16. Liu, B., Haftka, R.T., Akgum, M.A. Permutation genetic algorithm for stacking sequence optimization, AIAA-98-1830, 1141–1152 (1998)
17. MIL-S8698, Structural design requirements, helicopters, 3.1.10 Load Factor
18. Johnson, W.: CAMRAD II Comprehensive analytical model of rotorcraft aerodynamics and dynamics. Johnson Aeronautics (2008)
19. Hodges, D.H., Popescu, B.: On asymptotically correct timoshenko-like anisotropic beam theory. Int. J. Solids Struct. **37**, 535–558 (1999)
20. Cesnik, C.E.S., Hodges, D.H.: Variational-asymptotical analysis of initially twisted and curved composite beams. Finite Elem. Anal. Des. **1**, 177–187 (1994)
21. Hodges, D.H., Yu, W.: A rigorous, engineer-friendly approach for modelling realistic, composite rotor blades. Wind Energy **10**, 179–193 (2007)
22. Volovoi, V.V., Hodges, D.H., Cesnik, C.E.S., Popescu, B.: Assessment of beam modeling methods for rotor blade applications. Math. Comput. Model **33**, 1099–1112 (2001)
23. Pilkey, W.D., Wunderlich. W.: Mechanics of structures variational and computational methods (1994)
24. Librescu, L., Song, O.: Thin-walled composite beams theory and application (2006)
25. Hodges, D.H., Popescu, B.: Asymptotic treatment of the trapeze effect in finite element cross-sectional analysis of composite beams. Int. J. Nonlinear Mechanics **34**, 709–721 (1999)
26. Hodges, D.H.: A mixed variational formulation based on exact intrinsic equations for dynamics of moving beams. Int. J. Solids and Structures **26**, 1253–1273 (1990)
27. Patil, M.J., Althoff, M, Energy-consistent, galerkin approach for the nonlinear dynamics of beams using mixed, intrinsic equations, 47th AIAA Conference (2006)
28. Johnson, E.R., Vasiliev, V.V., Vasiliev, D.V.: Anisotropic thin-walled beams with closed cross-sectional contours. AIAA **39**, 27–38 (1998)
29. Ball, N.R., Sargent, P.M., Ige, D.O. Genetic algorithm representations for laminate layups. Artif. Intell. Eng. Des. Anal. Manuf. **8** 99–108 (1993)
30. Conceição António, C., Hoffbauer, L.N.: Uncertainty analysis based on sensitivity applied to angle-ply composite structures. Reliab. Eng. Syst. Safety **92**, 1353–1362 (2007)
31. Keller, D.: Optimization of ply angles in laminated composite structures by a hybrid, asynchronous, parallel evolutionary algorithm. Composite Structures **92**, 2781–2790 (2010)

A Finite Element Approach for the Vibration of Single-Walled Carbon Nanotubes

Seyyed Mohammad Hasheminia and Jalil Rezaeepazhand

Abstract We worked on vibrational aspects of zigzag single wall nanotubes using the finite element simulation. We modeled the nanotube and graphene sheet as 3D frame structures using beam elements. The natural frequencies of vibration and their vibration modes are obtained. The simulations are done for the zigzag nanotubes (8, 0). The first five natural frequencies are obtained for aspect ratios in the range of 4 to 20. The results show that the natural frequencies decrease as the aspect ratios increase. The results also indicate similar trends with results of previous studies for CNTs using universal force field potential. We also present frequencies of nanotube versus aspect ratio of SWNT in two boundary conditions (clamped–clamped and clamped-free), and compare it with solutions taken from MD, local shell model and nonlocal shell model which still remains the same trends and are close enough.

Keywords Carbon nanotubes · Vibration analysis · Beam element · Finite element · SWNT

1 Introduction

Significant researches have been done in the area of nano science and technology in the past two decades. As one of the most interesting nanomaterials, carbon nanotubes (CNT) have received huge attention in terms of fundamental properties

S. M. Hasheminia (✉)
Department of Mechanical Engineering, Ferdowsi University of Mashhad, No. 653 Vakil abad Blvd, After danesh Amooz St, 91889-93385 Mashhad, Iran
e-mail: mohamad.hasheminia@gmail.com

J. Rezaeepazhand
Department of Mechanical Engineering, Ferdowsi University of Mashhad, No. 653 Vakil abad Blvd, After danesh Amooz St, 91775-1111 Mashhad, Iran
e-mail: jrezaeep@um.ac.ir

and applications. This is largely because of significant physical properties found from both theoretical and experimental studies. For example, the electrical properties of CNT may become regulated by mechanical deformation. The deformation and vibration aspects of CNTs have been the subject of many experimental, molecular dynamics analysis (MD) and continuum modeling projects. Experiments at nano scale are still being improved in many cases and thus led us to valuable mechanical properties and helped to simulate numerous engineering problems. These properties are of real interest for applications such as sensors and smart materials. The study of these aspects is multi-disciplinary and involves various branches of science and technology. Steady approach has been made in exploring the mechanical properties of two types of CNTs: single-walled carbon nanotubes (SWCNT) and multi-walled carbon nanotubes (MWCNT). The measured specific tensile strength of a single walled carbon nanotube can be as high as 100 times that of steel, and the graphene sheet (in-plane) is as stiff as diamond at low strain. These significant mechanical properties motivate future study of possible applications for lightweight and high strength materials. Composite materials made of SWCNT or MWCNT have been fabricated and considerable enhancement in mechanical properties has been recently reported [1, 2].

As dealing with experiments at the nanoscale is difficult, theoretical analyses of nanomaterials become very important. Two basically different approaches are available for theoretical modeling of nanostructured materials: the atomistic approaches and the continuum mechanics. The first one includes the classical molecular dynamics (MD) [3–7]. MD approaches are often expensive to compute, especially for large-scale MWNTs. Hence, the continuum mechanics is increasingly being considered as an alternative way of modeling materials at the nanometer scale. In the classical (local) continuum models, CNTs are taken as linear elastic thin shells [8, 9]. It is concluded that the applicability of classical continuum and finite element models at very small scales is questionable. Therefore, continuum models need to be further extended to consider the size effects in nanomaterials studies. Many attempts have been made to develop more sophisticated types of continuum models in order to better accommodate the results from the MD simulations.

2 Brief Review of the Continuum Approach for Nanotube Analysis

Ru concentrated on buckling of MWNTs, and presented that the critical strain can be overestimated in certain cases if the VDW (van der Waals) forces aren't strong enough. In this situation, each single tube acts independently, and the smallest diameter tube fails first; and the others fails sequently. Actual bending stiffness of carbon SWNTs is low—about 25 times lower than that predicted by the elastic shell model if a representative thickness of 0.34 nm is used. However, there is the possibility of using MWNTs to improve bending strength In addition, they used an Elastic model to study buckling of a double-walled CNT under compression. It could be shown that CET (classical elasticity theory) works well, but existing CET models cannot be used directly for CNTs, because of the lack of study of a vdW interaction. They use the Airy stress function. It was found that adding an inner tube cannot increase the critical axial strain, however it can increase the force because of changing in cross-section [10, 11].

Govindjee and Sackman used the equations of continuum beam theory to interpret the mechanical response of NTs, specifically, using the Bernoulli-Euler equation for the Young's modulus. Reports that show an explicit dependence of "material" properties on system size when a continuum cross-section is assumed. They reported that the super-high values of modulus reported are because of the breakdown of the continuum hypothesis. Other modes can be used to define E, and we should expect that the value of E is not dependent on the mode of deformation. If E is different in tension and compression, then you have to look at how this would affect the bending stiffness [12].

Yakobson, Campbell, Brabec and Bernholc considered the way that intrinsic symmetry of a graphite sheet is hexagonal, and a 2D hexagonal structure is isotropic. Thus, it can be represented by a uniform shell with only 2 elastic parameters: resistance to in-plane bending (in-plane stiffness C) and flexural rigidity D. They quote another reference to get $C = 59$ eV/atom $= 360$ J/m^2, and $D = 0.85$ eV. They estimate the Poisson's ratio to be 0.19 based on simulation, and this corresponds to the experimental value for single crystal graphite. Claims that using $Y = 5.5$ TPa and $h = 0.066$ nm gives the correct values for the modulus and flexural rigidity. Choosing a more "natural" value for h (say 0.34 nm, the graphite interlayer spacing) really overshoots the rigidity. It was found that is useful to use these values for the shell continuum model. They estimate that the bulk modulus is slightly higher than diamond, and significantly higher than graphite. Also with properly chosen parameters, the continuum shell model provides a remarkably accurate "roadmap" of NT behavior beyond Hooke's law. They model the NT using a realistic many-body Tersoff-Brenner potential and MD, and then show that these transformations can be modeled with a continuous tubule model. Using this, they can model the behavior beyond the linear response [4, 13].

3 Finite Element Model of the Nanotube

We modeled the nanotube and graphene sheet as 3D frame structures using beam element. In this case carbon atoms are simulated as the nodes and the C–C bonding as the beam element. Due to this idea the molecular dynamics approach is linked with continuum mechanics. To reach a validate point of view we made the 3D structure as mentioned and assumed to be as a nanotube filament with definite geometry that means NT wall thickness and cross section. It was also assumed that the NT's global properties are valid for this structure. Considering definite beam properties we can draw a stress–strain graph for our model.

4 Vibration Properties of the Model

The vibrational aspects of CNTs have been the focus of considerable research. Many of the studies have been conducted based on the classical continuum mechanics including the Bernoulli–Euler/Timoshenko beam models, shell models and space truss/frame models. Most of the previous nonlocal vibration articles on CNTs have been established on the basis of the nonlocal Bernoulli–Euler/Timoshenko beam models [14]. By assuming proper Young's modulus for the NT model and deriving proper properties for the beam elements we investigate the vibrational properties of zigzag single-wall carbon nanotubes (CNTs) using FEM simulation in Ansys. The natural frequencies of vibration and their associated intrinsic vibration modes are obtained. The simulations are done for the zigzag nanotube (8,0). The first five natural frequencies are obtained for aspect ratios ranging from 4 to 20.

The resonant frequencies of SWCNT depend on the tube diameter and length. The atomic structures of SWCNTs could also exert important influence on their vibration behaviours. Thus, in this work, we analyze one group of SWCNT resonators, i.e., for zigzag nanotube (8,0). Chowdhury presented natural frequencies of four types of zigzag nanotubes (5,0), (6,0), (8,0) and (10,0) using universal force field potential [15].

Arash and Ansari presented frequencies of nanotube versus aspect ratio of SWNT in two boundary conditions (clamped–clamped and clamped-free), using three solutions MD, local shell model and nonlocal shell model [14].

5 Results and Discussion

The vibration frequencies of SWCNT depend on the length. The atomic structures of SWCNTs could also exert significant influence on their vibration aspects. Thus, in this work, we analyze just a zigzag nanotube (8,0). The computational results of the first five vibrational frequencies of the zigzag SWCNT are shown in Fig. 1, respectively, as functions of the length-to-diameter aspect ratio.

We also considered different boundary conditions. Our results are shown in Figs. 2, 3, 4 and 5, which are in good agreement with previous studies.

As shown in Fig. 1, for SWCNTs with the aspect ratio rising from around 4 to 20, the fundamental frequencies are in the ranges of 120–2500 GHz for the zigzag CNTs (8,0). It may be noted from the mentioned references. that the resonant frequencies of SWCNTs obtained based on the present MM approach are close to, but lower than, those given by a structural mechanics approach. Specifically, the

A Finite Element Approach for the Vibration 143

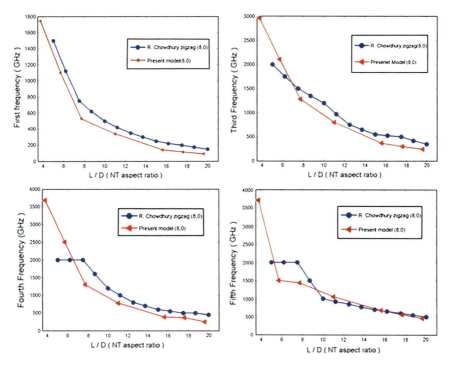

Fig. 1 First five vibrational frequencies of zigzag CNTs as a function of tube aspect ratio (L/D). First two modes are identical with different symmetric planes of flexural vibration. The 3rd mode corresponds to torsional vibration. The 4th and 5th modes correspond to higher flexural vibration

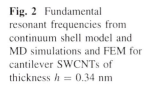

Fig. 2 Fundamental resonant frequencies from continuum shell model and MD simulations and FEM for cantilever SWCNTs of thickness $h = 0.34$ nm

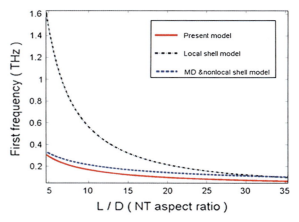

tendency of frequency to change with aspect ratio are generally in agreement with that given by these studies. For zigzag SWCNT, the frequencies of all five modes are generally in decreasing trend when the aspect ratio increases from 4 to up to

Fig. 3 Fundamental resonant frequencies from continuum shell model and MD simulations and FEM model for clamped SWCNTs of thickness $h = 0.34$ nm

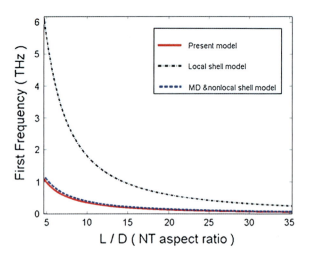

Fig. 4 Third resonant frequencies from continuum shell model and MD simulations and FEM model for cantilever SWCNTs of thickness $h = 0.34$ nm

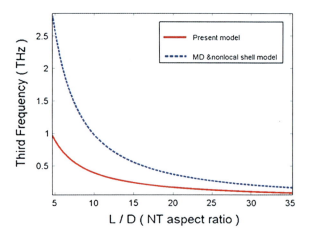

Fig. 5 Third resonant frequencies from continuum shell model and MD simulations and FEM model for clamped SWCNTs of thickness $h = 0.34$ nm

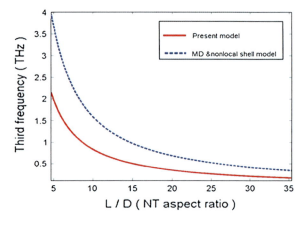

20. As we can see by increasing aspect ratio we have closer results in all approaches that gives the idea of considering long NTs as a general beam rather than shell.

6 Conclusions

The vibrational aspects of zigzag single-wall carbon nanotubes are studied in this research. A continuum mechanics based approach is used to estimate the frequencies. In this study, we used a finite element model shaped as a 3D frame structure to be our NT and used beam elements as C–C bondings. Vibration of CNTs show features of decreasing frequencies with increase in aspect ratio. It is found that, natural frequencies of zigzag CNTs are highly depended to aspect ratio specially in short NTs. However, differences in frequencies vanishes with increase in aspect ratio. In view of results presented, it follows that the frequency of SWCNTs is primarily determined by their geometric sizes, i.e., diameter and the aspect ratio, but cannot be substantially changed due to the variation of their atomic structures.

References

1. Mamedv, A.A., Kotov, N.A., Prato, M., Guldi, D.M., Wicksted, J.P., Hirsch, A.: Molecular design of strong single-wall carbonnanotube/polyelectrolyte multilayer composites. Nat. Mater. **1**, 190–194 (2002)
2. Arroyo, M., Belytschko, T.: Continuum Mechanics Modelling and Simulation of Carbon Nanotubes. Kluwer Academic Publishers, Dordrecht (2005)
3. Iijima, S., Brabec, C., Maiti, A., Bernholc, A., Chem, J.: Structural flexibility of carbon nanotubes. Phys **104**(5), 2089–2092 (1996)
4. Yakobson, B.I., Campbell, M.P., Brabec, C.J., Bernholc, J.: Comput. Mater. Sci. **8**, 341–348 (1997)
5. Hernandez, E., Goze, C., Bernier, P., Rubio, A.: Elastic properties of C and BxCyNz composite nanotubes. Phys. Rev. Lett. **80**, 4502–4505 (1998)
6. Sanchez-Portal, D., Artacho, E., Soler, J.M., Rubio, A., Ordejon, P.: Ab initio structural, elastic, and vibrational properties of carbon nanotubes. Phys. Rev. B **59**, 12678 (1999)
7. Qian, D., Wagner, J.G., Liu, W.K., Yu, M.F., Ruoff, R.S.: Mechanics of carbon nanotubes. Appl. Mech. Rev. **55**, 495–553 (2002)
8. Yakobson, B.I., Brabec, C.J., Bernholc, J.: Nanomechanics of carbon tubes: instabilities beyond linear response. Phys. Rev. Lett. **76**, 2511–2514 (1996)
9. Ru, C.Q., Mech, J.: Axially compressed buckling of a double-walled carbon nanotube embedded in an elastic medium. Phys. Solids **49**, 1265–1279 (2001)
10. Ru, C.: Column buckling of multiwalled carbon nanotubes with interlayer radial displacements. Phys. Rev. B **62**(24), 16962–16967 (2000)
11. Ru, C.: Effect of van der Waals forces on axial buckling of a double-walled nanotube. J. Appl. Phys. **87**(10), 7227–7231 (2000)
12. Govindjee, S., Sackman, J.: On the use of continuum mechanics to estimate the properties of nanotubes. Solid State Commun. **111**, 227–230 (1999)

13. Yakobson, B., Brabec, C., Bernholc, J.: Nanomechanics of carbon tubes: instabilities beyond linear response. Phys. Rev. Lett. **76**(14), 2511–2514 (1996)
14. Arash, B., Ansari, R.: Evaluation of nonlocal parameter in the vibrations of single-walled carbon nanotubes with initial strain. Physica E **42**, 2058–2064 (2010)
15. Chowdhury, R., Adhikari, S., Wanga, C.Y., Scarpa, F.: A molecular mechanics approach for the vibration of single-walled carbon nanotubes. Comput. Mater. Sci. **48**, 730–735 (2010)

Characteristics of Welded Thin Sheet AZ31 Magnesium Alloy

Mahadzir Ishak, Kazuhiko Yamasaki and Katsuhiro Maekawa

Abstract Conventional arc welding processes are difficult to use to join thin sheet magnesium alloy because of the necessity of high energy input, which in turn leads to various problems such as burn through and distortion. Alternatively, laser welding can resolve these problems because of lower heat input and smaller spot size compared to conventional welding. Even when using laser welding, it is difficult to weld thin magnesium sheets with a thickness of less than 1 mm; cut, melt through and cracks tend to occur due to the evaporation of molten metal and high solidification rate. In this study, an attempt has been made to lap fillet welding of thin sheet magnesium alloy AZ31B with a thickness of 0.3 mm using a pulsed Nd:YAG laser beam in a conduction mode. This paper investigates the occurrence of defects in the lap fillet joint of AZ31B magnesium alloys. Defects such as void and cracks were observed at the weld root. A void at the root occurred because of lack of fusion due to insufficient melting of the lower sheet. The void was reduced by grinding the metal surface to eliminate the oxide layer. Cracks generated in large grain areas initiated from the void at the root. A higher scan speed significantly improves the defect behaviour because of generating a narrow large grain area and wider fine grain area. Macropore-free weld was obtained in this laser welding research, and smaller amount of micropores than the base metal can be attained.

Keywords Laser Welding · Thin sheet · Magnesium alloy · Welding defects

---

M. Ishak (✉)
Faculty of Mechanical Engineering, University Malaysia Pahang, 26600 Pekan, Pahang
e-mail: mahadzir@ump.edu.my

K. Yamasaki · K. Maekawa
The Research Center for Superplasticity, Ibaraki University, Ibaraki, Japan
e-mail: kyama@mx.ibaraki.ac.jp

K. Maekawa
e-mail: mae@mx.ibaraki.ac.jp

1 Introduction

Welding of thin sheets usually experiences many problems compared with that of thick sheet metal. In the conventional arc welding processes, a high heat input usually causes various problems such as cutting, burn through, distortion, porosity, cracking, etc. Thus, the utilization of the proper welding process, procedure and technique is important in order to deal with these problems. Comparing with arc welding, laser welding and electron-beam welding have many advantages, such as narrow weld and high penetration depth. Laser beam is preferable because it can be used in ambient pressure and temperature rather than electron beam welding. Even with many advantages, laser welding of thin sheet metals still has problems, including the loss of material due to evaporation and improper control of heat which leads to the formation of cut and melt-through. There are many reports regarding welding of copper, stainless steel and aluminium alloys of thin sheets with a thickness of less than 1 mm [1–3], but few studies focusing on thin sheet at similar thickness of magnesium alloys welding have been carried out.

In this paper, a Nd:YAG laser was utilized to joint thin sheet AZ31B magnesium alloys. The rationale for the selection of the thin sheet AZ31B magnesium alloys was because of its unique properties; e.g. low density, high strength-to-weight ratio, high damping capacity and good recyclability. Thus, recently an increased attention has gained in many industries [4]. However, processing of magnesium alloys is difficult because cracks, pores and cuts easily generate [5–7]. In this research, laser welding with a conduction mode rather than a keyhole one is employed, because a very stable weld pool could be provided. The absence of unstable fluid motion, which usually takes place in a keyhole mode welding, could give attractive weld quality as well as good controllable penetration depth. These advantages of conduction-mode welding are much more suitable to weld thin sheet magnesium alloys as thin as 0.3 mm.

Therefore, the objective of this research is to investigate and understand the mechanism of the formation of defects in lap fillet joint of the AZ31B magnesium alloy with a thickness of 0.3 mm by using a pulsed Nd:YAG laser.

2 Experiment

Lap fillet welding by laser was performed on a 0.3 mm thick magnesium alloy sheet AZ31B. The size of the sheet is 20 mm × 30 mm. The composition of the magnesium alloy is listed in Table 1. Before welding, the sheets were ground with SiC paper (#1000), and then cleaned with ethanol to remove oxide films and oil. A

Table 1 Composition of magnesium alloy AZ31B (mass %)

Material	Al	Zn	Mn	Mg
AZ31B	2.63	0.28	0.71	Bal.

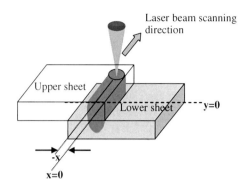

Fig. 1 Beam centre location defined from edge of upper sheet

pulsed Nd:YAG laser with a 1.06 μm wavelength, 0.4 mm beam spot diameter and 100 mm focal length lens was used for this experiment. Figure 1 illustrates the setup for the lap fillet welding experiment. Lap welding was carried out by overlapping two sheets, and then joined by the laser at the edge side of the upper sheet. A specimen was fixed with clamps and jigs in a metal closed box, and then located on a CNC X–Y table. The top of the box was closed by a heat resistance glass, through which the laser beam is transmitted during welding. Argon gas was continuously flowing inside the box at a rate of 20 l/min during the process.

In this experiment, the two sheets were welded with a gap width of less than 35 μm. The laser beam centre location (x) was varied from the edge of the upper sheet towards the –x direction as shown in Fig. 2: 0.1, 0, −0.1, −0.2 and −0.3 mm with various scan speed from 50 to 600 mm/min. The setting of the laser beam on a specific location on the upper sheet was determined by a camera and the precisely controllable CNC table within an accuracy of 1 μm. Other parameters of the Nd:YAG laser welding such as pulse energy, pulse duration, and repetition rate were fixed at 1.8 J, 3.0 ms and 80 Hz, respectively.

The laser was scanned in the y-direction for welding. The welded specimens were cross-sectioned, ground and polished. The mounted specimens were etched in a solution composed of either 10 ml acetic acid diluted with 100 ml distilled water for macroscopic observation, or solution of 5 ml acetic acid +5 g picric acid +10 ml distilled water, and 70 ml ethyl alcohol for microstructure observation. The macrostructure and microstructure were characterized by optical microscopy (OM) and scanning electron microscope (SEM). Elements in the crack area were determined by EDX analysis.

3 Results and Discussion

3.1 Weld Geometry

To analyse the weld geometry of a welded sample, the average weld width and penetration depth were measured. The effects of the centre beam location on the

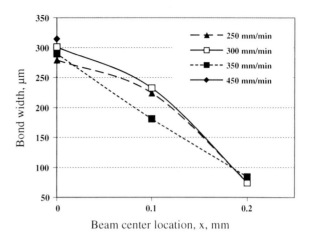

Fig. 2 Effect of beam location on bond width

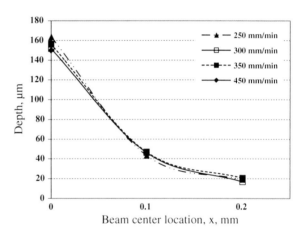

Fig. 3 Effect of beam location on penetration depth

bond width and penetration depth are shown in Figs. 2 and 3. Both of them decrease significantly with increasing the distance of beam centre location from the edge of the upper specimen. A large distance of beam centre location results in a low concentration of heat on the upper surface of the specimen. This happened due to heat dissipation in the magnesium alloy having high thermal conductivity. Therefore, less heat was absorbed by the workpiece resulting in a smaller melted weld area. Increases in distance of beam centre location could also enlarge the size of voids at the root as shown in Fig. 4. A detailed description of the void formation at the root will be provided in Sect. 3.3. From this result, we could understand that the weld geometry in laser welding is very sensitive to the beam centre location distance in the order of 100 μm. Thus, in this paper we fix the beam centre location on the upper edge ($x = 0$) to optimize the weld geometry and reduce defects.

Fig. 4 Weld cross-section at beam location x = −0.2 mm at scan speed 250 mm/min

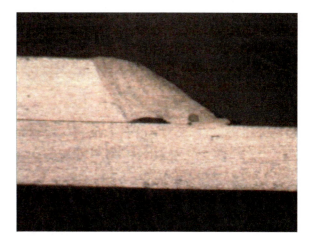

3.2 Microstructure

Figure 5 features the microstructure of the weld zone: (a) shows the upper weld at 450 mm/min, which consists of fine and equiaxed grains. A low scan speed of 250 mm/min results in larger grains growth at the weld centre as shown in Fig. 5b. This may be caused by the high heat input and then the longer cooling time of the molten metal. Deeper location generates large grain, especially at the boundary of the base metal as shown in Fig. 5c. It can be observed that the elongated cellular grains grow perpendicular to the fusion boundary. In other words, grains coarsen with decreasing distance from the weld centre because the thermal cycle changes. The area with long and elongated grains was narrower as scan speed increases, in which the length at a scan speed of 450 mm/min is about 80 to 100 μm. The existence of this zone stems from the physical properties of the AZ31B magnesium alloy: a high thermal conductivity and a low thermal capacity. In addition, note that a conduction mode was applied rather than a keyhole mode in the present study.

Some black pits and white dots can be observed in the fine precipitated inclusions at the weld area as shown in Fig. 5a. EDX analysis revealed that the black pits contain high Al, whereas the white elements contain high Al and Zn compared with the base metal. In general, these participates have been identified as $Mg_{17}Al_{12}$ and $Mg_{17}(Al,Zn)_{12}$ [8, 9].

3.3 Defects

3.3.1 Voids

Voids or discontinuity at the weld root occurred likely because of lack of fusion [10–13]. As shown in Fig. 6, it seems that this happens because of insufficient

Fig. 5 Microstructure of (**a**) upper weld at scan speed 450 mm/min (**b**) upper weld at scan speed 250 mm/min and (**c**) fusion boundary at scan speed 450 mm/min

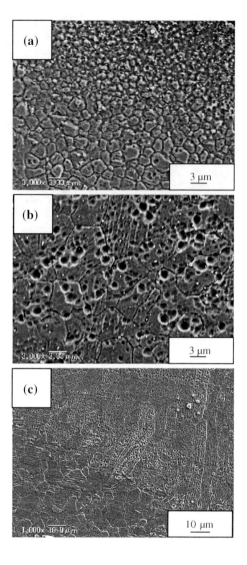

melting of the lower sheet although the upper sheet was heated at the welded root area. Insufficient melting is caused by inadequate energy input and welds preparation at the welded root area [10–13]. The existence of a narrow gap could also cause the lack of fusion occurred. Furthermore, it is plausible that the oxide thin film formed on the surface of the lower sheet, although it was ground and cleaned. Magnesium is an active metal, in which oxide could easily form on the surface when in contact with air after grinding [14]. This thin layer could act as a barrier that prevents heat transfer to the lower sheet.

A thin sheet without surface grinding on the upper surface of the lower sheet and the upper sheet with grinding and cleaning was welded at a scan speed of

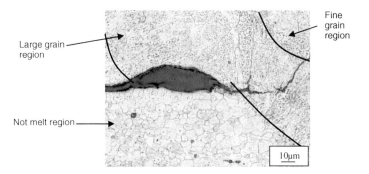

Fig. 6 Microstructure around void at root

350 mm/min. The cross-section images of the weld with and without grinding/cleaning are shown in Fig. 7a, b, respectively. Large voids and discontinuities are observed when the sample was not ground, and the melting area is much smaller compared with the ground welded sample. This is because a thick oxide layer exists on a surface of the lower sheet, acting as a heat insulator. This plays a significant role on the weld, especially in this conduction mode laser welding process. In addition, the internal stress produced during solidification and cooling results in the separation of the surfaces sticking with each other, thus causing the expansion of a void or discontinuity as shown in Fig. 6. Although a new oxide instantaneously forms on magnesium in air after polishing/cleaning, the newly formed oxide layer was expected to be much thinner, so that the formation of the defect is reduced. This implies that the oxide film had a large effect on the occurrence of voids or discontinuities at the root. The defects severely happened at low scan speeds but significantly improved at higher scan speed. Higher scan speeds result in low heat input with less stress during melting and solidification. Besides, high scan speeds increase the finer grain area and reduce the large grain area. Therefore, these features characterize the improvement of the weld defect at the root in high scan speeds.

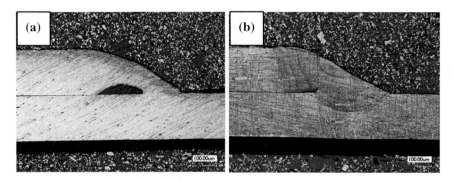

Fig. 7 Cross-section of welded samples with different surface preparation (**a**) lower sheet not polished but only cleaned and (**b**) lower sheet polished and cleaned

3.3.2 Cracks

Cracks can be seen at low speeds when x = 0, and become much smaller as the scan speed is increased. No visible cracks were observed at the beam center location x = 0 with scan speeds at 400 and 450 mm/min. Furthermore, no macro cracks were observed on the surface of the upper sheet weld bead. As can be seen in Fig. 6 cracking initiated from the root and then propagated into the weld metal, but not to the center or the upper surface of the weld zone. From a microstructure image of Fig. 8, cracking occurred and propagated in the large grain region in a transgranular manner.

The crack observed in tungsten inert gas welding of AZ91D is related to the liquation of a second phase or low melting-point precipitates in the heat affected zone (HAZ) or partially melted zone (PMZ) [15]. This type of crack was a "liquation cracking" and is due to the low melting point of the intermetallic compound at the grain boundary of the HAZ and PMZ, which greatly decreases the strength of the weld [15, 16]. Figure 8 also shows the result of EDX line analysis of cracks connecting to the void formed at the root. No peaks of Al, Mn and Zn are detected at open surface of cracks. Accordingly, there was no segregation of Zn, Al and Mn at this area, meaning that, the second phase of low-melting intermetallic compounds consisted of Mg and Al or Zn do not precipitate near the crack area. No finer grains were observed at the open crack, which is inconsistent with references [15, 16]. No significant change of hardness was

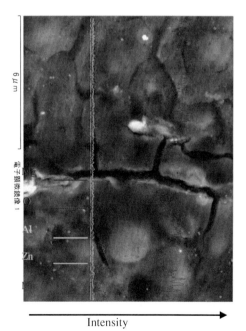

Fig. 8 EDX line analysis of crack near to void

detected at this area [17]. Consequently, no precipitation of second phase low melting intermetallic compounds took place at the crack region.

However, the peak of oxygen is observed near to the crack open surface. Although the surface of the thin sheet was ground with SiC paper prior to welding, it is plausible that a thin oxide had already been formed because of high the chemical activity of magnesium. On the other hand, it is thought that the crack initiation from a void at the root is due to high stress concentration, followed by propagation to the large elongated grain area along the thin oxide film. During low scan speed welding, wider area of large grains are formed at the boundary of the fusion zone and the base metal [17]. This area is exposed to much lower temperature than at the middle of fusion zone, so that the oxide film is not properly broken or melt during welding. Furthermore, low scan speed results in high heat input which consequently produces higher stress in the large grain region in the course of metal solidification and contraction. Therefore, the cracks easily propagate through the oxide film area.

Higher scan speeds tend to reduce stress as well as lead to the formation of a wider fine equiaxed grain region, so that the large grain region becomes much smaller, as shown in Figs. 5a, c. These fine equiaxed grains are less susceptible to cracking than larger ones because the stress is more evenly distributed among numerous grain boundaries.

3.3.3 Porosity

Pores are detrimental to welding part. Thus, the porosity in the weld zone was measured in total area of 40×30 μm^2 by using a SEM and measurement software. Figure 9 shows the change of porosity area percentage and the number of pores with varying scan speed. The porosity area percentage decreases significantly from 19.4 to 1.2 % as the scan speed increases from 250 mm/min to 450 mm/min. A similar tendency can be seen between the scan speed and the number of pores.

Figure 10 shows the relationship between the distribution of average size and the number of pores at different scan speed. The majority of pore sizes are smaller than 0.4 μm, while the highest number of pores has a size of around 0.25 μm. There are two types of porosity produced when welding non ferrous metal: macro and micropores. Macropores are larger than 0.2 mm, while micropores are as large as several micrometers [18]. From Figs. 9 and 10, that macro porosity-free welds can be obtained using a pulsed Nd:YAG laser at all scan speeds.

One of the important reasons for the formation of pores in laser welding of magnesium alloys is the keyhole instability. The collapse of the unstable keyholes and turbulent flow of molten metal during keyhole mode welding may cause the formation of porosity in the magnesium alloy [19, 20].

In the present research, however weld is featured by the conduction mode rather than keyhole one. Thus, the keyhole instability was not a main reason of pore formation. In addition, it is reported that pores are formed in AZ31B welding using

Fig. 9 Porosity and number of pores at different scan speed

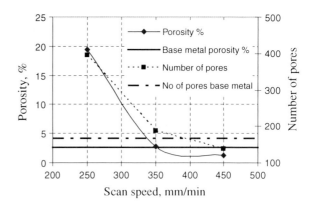

Fig. 10 Pore size distribution

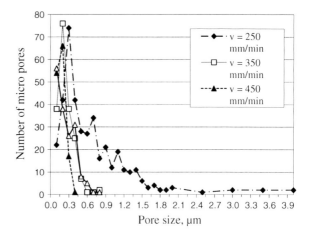

hybrid laser–tungsten because of lack of shielding, where air is trapped into the molten pool during welding [21]. In our case welding was carried out in a closed box with argon gas environment, so that this kind of pores could be prevented. From Fig. 9, the porosity at a scan speed of 250 mm/min is significantly high compared with that of the base metal. The number of pores at this scan speed is also high, and the size of pores is distributed widely. It can be seen from Fig. 10 that the average micropores size at this scan speed is 2.5 times higher than that of the base metal: the content of pores with larger than 1 µm is around 20 %. These pores are most probably caused by expansion and coalescence of pre-existing pores during welding. The coalescence and expansion of small pre-existing pores stemmed from heating as well as reduction of internal pressure which lead to a high percentage of pores in the weld zone [20]. Low scan speed causes high heat input and much time available for expansion and coalescences of many pre-existing micropores. Micropores with larger than 1 µm are not observed at high scan speeds, because of short time for the generation of expansion and coalescence of pre-existing.

However, coalescence and expansion of pre-existing pores alone cannot explain the increase of porosity and the number of micropores in the weld zone at a low scan speed of 250 mm/min. The number of pores with size smaller than 1 μm is dominant in the weld zone. These pores are most probably caused by hydrogen rejection from the solid metal. It is reported that the rejection of hydrogen from the solid–liquid assists in nucleation and growth of micro-porosity during solidification [19–21]. The liberation of hydrogen gas released during the solidification of the Al causes the formation of micropores in the case of tungsten arc welding of AZ91D [21]. A similar phenomenon is likely to occur when AZ31B was laser-weld at a very low scan speed. Increase in the scan speed prevents the formation of micropores because a high cooling rate results in less time for hydrogen to diffuse [18]. Thus, it is assumed that the role of hydrogen on porosity formation is insignificant during laser welding of AZ31B thin sheet at high scan speeds above 350 mm/min.

The porosity and the number of pores of the welded sample at high scan speeds are slightly lower than in the base metal. It is thought that some amounts of pores were released during heating to reduce the porosity and the number of pores. Therefore, it can be concluded that the scan speed is a key factor to prevent the formation of micropores in the thin sheet laser welding.

3.4 Tensile Test

The tensile fracture loads of the welded joints with varying beam location and scan speed are shown in Fig. 11. This result indicates that the fracture load increases with decreasing beam centre location. In addition, increase in scan speed results in stronger weld. These phenomena can be explained by the result of macro and microstructure observations. The increase of the fracture load contributes to the improvement of bond width (Fig. 12) as the beam centre location varies from

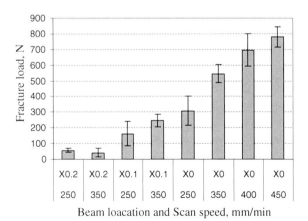

Fig. 11 Fracture load at different beam location and scan speed

Fig. 12 Schematic illustration of bond width and throat length

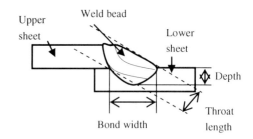

Fig. 13 Throat length at different scan speed

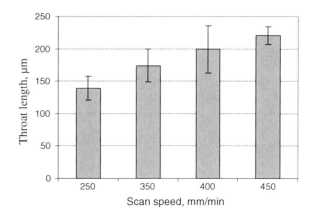

x = −0.2 mm to 0. Large distance of the beam centre from the upper edge produces a small molten pool because of low heat concentration at the upper specimen. Nevertheless, fracture load improvement is significant as scan speed increases at the beam centre location x = 0 in spite of minimum improvement in bond width. The improvement of the fracture load at x = 0 is primarily because of the enhancement of throat length rather than bond width as shown in Fig. 13. Increase in throat length is dependent on the size of the defect that occurs at the weld root. Wider throat length at high scan speed is closely related to small size of void at the weld root.

At low scan speeds below 350 mm/min, high porosity and serious cracks in the weld bead are formed. These defects all negatively affect the tensile properties of the joints. In addition, the grain fusion zone becomes coarser with decreasing scan speed, which also decreases the tensile properties. Therefore, in order to provide less defects and higher tensile properties of welded joints, it is recommended that the beam centre should be located at the edge and high scan speeds be set at above 350 mm/min.

4 Conclusions

A pulsed Nd:YAG laser system with a conduction mode has been used to weld 0.3 mm fillet lap joints made of the wrought magnesium alloy AZ31B. This study yields the following results:

(1) Voids and discontinuity at the root occur because of insufficient melting. A narrow gap and oxide layer on the lower surface act as a heat barrier to prevent the melting of the lower sheet. The cracks generated at the weld are originated from a void at the root and propagated in a transgranular manner in the large grain area. Higher scan speed significantly reduces the defects, accompanied by narrower large grain area and wider fine grain region.
(2) Macropore-free weld has been obtained by this laser welding. Micropore porosity significantly reduced at high scans speed.
(3) Fracture load of the welded joints is significantly improved as the beam centre is located at $x = 0$ because of the resultant larger bond width. Besides, increase in the throat length stemmed from the improvement of defects at the weld root leads to higher fracture load at 400–450 mm/min.

References

1. Moon, J., Katayama, S., Mizutani, M., Matsunawa, A.: Lap welding characteristics of thin sheet metals with combined laser beams of different length. Jpn. Weld Soc. **20**, 468–476 (2002). (In Japanese)
2. Shizuo, U., Taisuke, A., Kanichiiro, S.: The welding conditions of very thin aluminum sheet of high welding. Jpn. Weld Soc. **11**, 361–364 (1993)
3. Leong, K.H., Sabo, K.R., Altshuller, B., Wilkinson, T.L., et al.: Laser beam welding of 5182 Aluminum alloy sheet. J. Las. A **11**(3), 109–118 (1999)
4. Aghios, E., Bronfiu, B., Eliezer, D.: The role of the magnesium industry in protecting the environment. J. Mater. Process Tech. **117**, 381–385 (2001)
5. Jinhong, Z., Lin, L., Zhu, L.: CO_2 and diode laser welding of AZ31 magnesium alloy. Appl. Surf. Sci. **247**, 300–306 (2005)
6. Toshikatsu, A., Hiroshi, T., Hitaka, I., et al.: Some characteristics of pulsed YAG laser welds of Magnesium Alloys. Nihon University research report no 38, 1–9 (2005)
7. Lung, K.P., Che, C.W., Ying, C.H., et al.: Optimization of Nd:YAG Laser welding onto magnesium alloy via Taguchi analysis. Opt. Laser Technol. **37**, 33–42 (2004)
8. Ghazanfar, A., Lin, L., Uzma, G., Zhu, L.: Effect of high power diode laser surface melting on wear resistance of magnesium alloys. Wear **260**, 175–180 (2006)
9. Quan, Y.J., Chen, Z.H., Gong, X.S., et al.: Effects of heat input on microstructure and tensile properties of laser welded magnesium alloy AZ31. Mater. Charact. **59**(10), 1491–1497 (2008)
10. Rihar, G., Uran, M.: Lack of fusion-Characterisation of indications. Weld World **50**(½), 35–39 (2006)
11. Abe, N., Tsukamoto, M., Morikawa, A., et al.: Welding of aluminum alloy with high power direct diode laser. Trans. JWRI **31**(2), 157–163 (2002)

12. Barsoum, Z., Lundback, A.: Simplified FE welding simulation of fillet welds-3D effects on the formation residual stresses. Eng. Fail. Anal. (2009). doi:10.1016/j.engfailanal.03018
13. Barsoum Z, Jonsson, B.: Fatigue assessment and LEFM analysis of cruciform joint fabricated with different welding processes. Inter Inst Welding Doc.no XIII-2175-07 (2007)
14. Teng-Shih, S., Jyun-Bo, L., Pai-Sheng, W.: Oxide films on magnesium and magnesium alloys. Mater. Chem. Phys. **104**, 497–504 (2007)
15. Zhou, W., Long, T.Z., Mark, C.K.: Hot cracking in tungsten inert gas welding of magnesium alloy AZ91D. Mat Sci and Tech **23**(11), 1294–1299 (2007)
16. Munitz, A., Cotler, C., Stern, A., et al.: Mechanical properties of gas tungsten arc welded magnesium AZ91D plates. Mat. Sci. Eng A **302**, 68–73 (2001)
17. Mahadzir, I., Yamasaki, K., Maekawa, K.: Lap fillet welding of thin sheet AZ31B magnesium alloy with pulsed Nd:YAG laser. J. Int. A Solids Mechanics **3**(9), 1045–1056 (2009)
18. Pastor, M., Zhao, H., Martukanitz, R.P. et al.: Porosity, underfill and magnesium loss during continuous wave Nd:YAG laser welding of thin plates of aluminum alloy 5182 and 5754. Weld Res. Supp 207–216s (1999)
19. Cao, X., Jahazi, M., Immarigeon, J.P., Wallace, W.: (2006) A review of laser welding techniques for magnesium alloys. J. Mat. Process. Tech. **171**, 188–204
20. Zhao, H., Debroy, T.: Pore formation during laser beam welding of die-cast magnesium alloy AM60b —mechanism and remedy. Weld Res Supp 204–210s (2001)
21. Jun, S., Guoqiang, Y., Siyuan, L., Fusheng, P.: Abnormal macropores formation during double-sided gas tungsten arc welding of magnesium AZ91D alloy. Mat. Charact. **59**, 1059–1065 (2008)

Localization of Rotating Sound Sources Using Time Domain Beamforming Code

Christian Maier, Wolfram Pannert and Winfried Waidmann

Abstract The motion of an acoustic source produces a Doppler shift of the source frequency which is dependent on the source's motion relative to the receiver. Some applications in acoustics involve rotating sound sources around a fixed axis in space. For example, the noise emitted by fans is of interest and because of the fast rotation, the sound sources are not easy to locate with the standard Delay-and-Sum Beamforming code. In the time domain approach for stationary sound sources, the Delay-and-Sum Beamforming works with shifting the microphone signals due to their different delays caused by the different distances between the source and the microphones and summing them up. This approach is adapted to a moving source, resulting in time dependent delays. The delays are calculated via an advanced time approach where the time at the receiver is calculated from the emission time τ plus a time dependent delay due to the time dependent distance $r(\tau)$. In contrast to the standard beamforming code, this time domain beamforming code allows to treat rotating sound sources as well as stationary sound sources. In this chapter the differences between the standard Delay-and-Sum Beamforming to the rotating time domain beamforming is shown and examples are presented.

Keywords Fan · Rotating sound sources · Retarded time · Beamforming

---

C. Maier (✉) · W. Pannert · W. Waidmann
Department of Mechanical Engineering, University of Applied
Sciences Aalen, Aalen, Germany
e-mail: Christian.Maier@htw-Aalen.de

W. Pannert
e-mail: Wolfram.Pannert@htw-Aalen.de

W. Waidmann
e-mail: Winfried.Waidmann@htw-Aalen.de

1 Introduction

The sound emitted by moving sound sources like, for example, a flowed airfoil or rotating fan blades are problems of technical interest. Visualising and analysing moving sound sources is much harder in comparison to stationary sound sources. The Doppler–shift and the retarded time due to the movement of the sound source have to be taken into account. In this work a time domain algorithm is presented which can be applied to rotating sound sources which are produced for example by fan blades. The theory is shown and the algorithm is proved with measurements using an acoustic camera at a rotating fan.

The standard Delay-and-Sum Beamforming method can be applied in time and frequency domain [1]. But the method is not suitable for moving sound sources. To compensate the movement of the sound source, special corrections are necessary. For this case the rotation of fans have to be compensated. The pressure field of a moving monopole is derived for a uniform flow in this approach. The approach presented below is original based on [2] and is based on the Delay-and-Sum Beamforming method in the time domain. A method to compensate rotating sound sources in the frequency domain, especially for high resolution beamforming techniques [3], is presented by Prof. Pannert from the University of Applied Sciences Aalen [4].

2 Theory

The movement of a point source can be treated with the Greens function approach for solving the inhomogeneous wave equation.

Taking the inhomogeneous wave equation for a stationary source located at $\vec{x}$

$$\frac{1}{c^2}\frac{\partial^2 p'}{\partial t^2} - \Delta p' = q(\vec{x}, t) \tag{1}$$

where $q(\vec{x}, t)$ is a the source distribution, the solution for free space conditions without boundaries for p' can be calculated from an integral formulation

$$p'(\vec{x}, t) = \frac{1}{4\pi} \int_{\mathbb{R}^3} \frac{q(\vec{y}, t - |\vec{x} - \vec{y}|/c)}{|\vec{x} - \vec{y}|} d^3\vec{y} \tag{2}$$

with

$$r = |\vec{x} - \vec{y}| \tag{3}$$

and the retarded time τ

$$\tau = t - |\vec{x} - \vec{y}|/c \tag{4}$$

The signal which was emitted at time τ at a position $\vec{y}$ and is observed at time t at the point $\vec{x}$. For the general source distribution a concrete source distribution can be inserted. The simplest model for a moving monopole is the distribution

$$q(\vec{x},t) = Q(t)\delta(\vec{x} - \vec{x}_s(t)) \qquad (5)$$

with $\vec{x}_s(t)$ as the actual time dependent position and $Q(t)$ as the amplitude of the monopole sound source.

Figure 1 shows the situation for a moving sound source and a fixed observer position. The observer point is the microphone position (at the microphone array).

It is necessary to calculate the distance between sound source and the microphone position for every time step τ_n to calculate the time delay to the observer position [5, 6].

In the retarded time approach, the retarded emission time τ is calculated back from the receiving time t via

$$\tau = t - r(\tau)/c \qquad (6)$$

and cannot be calculated analytically in general cases due to the complicated dependence of $r(\tau)$ from τ. It can numerically found as a root of Eq. (7) [6]. Algorithms that treat that problem can be found in [7] or [8].

In the advanced time approach which is applied in this work, the receiver time t can be calculated via

$$t = \tau + r(\tau)/c \qquad (7)$$

This is much easier, but results in unequally spaced time samples t_n when using equally spaced time samples τ_n.

In Fig. 2 the situation is shown for a moving source. An emitted signal at the time τ arrives at the observer position $\vec{x}$ at the time t. The speed of sound is c. In the case of a stationary source, the retarded time only depends on the position of

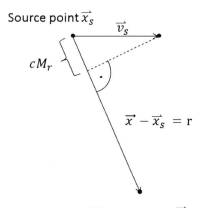

Fig. 1 Movement of the source term to a fix observation point

Fig. 2 Retarded time emitted from a moving sound source in the space–time

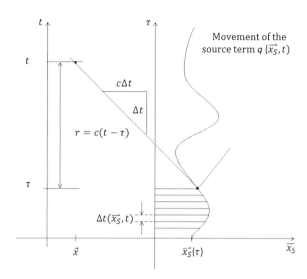

the source $\vec{x}_s$. In the case of a moving source it depends on $\vec{x}_s(\tau)$. This time delay is calculated for every time step n

$$\Delta t_n(\tau_n) = t_n - \tau_n = \frac{|\vec{x} - \vec{x}_s(\tau_n)|}{c}. \tag{8}$$

The condition $\tau_n < t_n$ is always fulfilled for the case that the source term moves with subsonic speed and $\Delta t_n(\tau_n)$ is always positive. Working with these time delays the motion of the source can be compensated in the received microphone signals and the moving source is imaged at a fixed position which corresponds to the position at time $\tau = 0$.

In Fig. 3, the typical set up for investigating a fan with an acoustic camera is shown. It is necessary to compensate the movement of the sound source. To compensate this movement, in this case the rotation of the fan with its blades, it is necessary to shift the time signals of every microphone for every time step at an amount, which is due to the change in distance between the moving source and the selected microphone. These shifted signals are used then to calculate the beam pattern.

In Fig. 4 simulated signals for a rotating source are shown. The pressure signal shows clearly the varying frequency due to the Doppler shift (Fig. 4a). The radial motion of the source is subsonic Fig. 4b shows the spectrum of the microphone signal in (a). The radius at which the sound source rotates is 0.65 m and the frequency of rotation is 100 Hz.

The frequency spectrum (Fig. 4b) shows the peak no longer at the position of 1,500 Hz. This is the effect due to the Doppler shift; the peak is now shifted in positive and negative frequency away from the emitted 1,500 Hz. Beside this, this

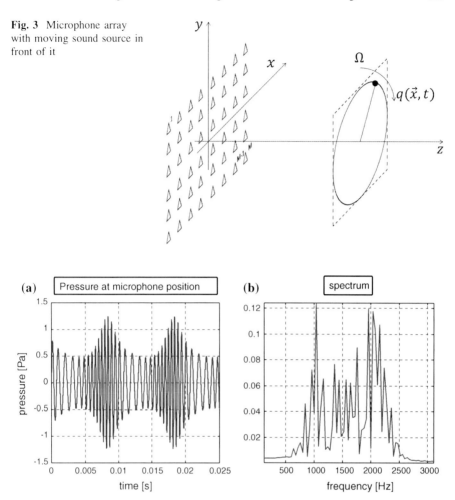

Fig. 3 Microphone array with moving sound source in front of it

Fig. 4 Received signal from a rotating source. Distance microphone–plane of rotation in distance $D = 1$ m; rotation speed $n = 6{,}000$/min; frequency of the source $f = 1{,}500$ Hz. **a** Microphone signal. **b** Spectrum of the pressure signal

frequency spectrum is not symmetric, because the motion between the source and the receiver also has an influence to the amplitude of the signal.

3 Measurements

To proof the programmed algorithm, first of all, Matlab tests were carried out. This Matlab code treats the theory explained in the last section. For this tests an artificial rotating sound source was simulated in Matlab, that rotates with 600 rpm on

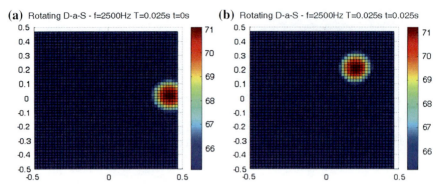

Fig. 5 Virtual Matlab sound source rotating anti clock wise around $x = 0.2$ m and $y = 0.0$ m, (**a**) at time 0 s and (**b**) at time 0.025 s

a circle with a radius of 0.3 m around the position $x = 0.2$ m and $y = 0.0$ m. The distance between the sound source and the microphone array is $D = 1$ m (Fig. 5).

4 Results

The algorithm used for the validation is implemented in the acoustic camera. So the analysis can be done and it is possible to compare it to the standard Delay-and-Sum Beamforming. Delay-and-Sum Beamforming only locates stationary sound sources and rotating beamforming locates moving sound sources. The Delay-and-

Fig. 6 Analysis of a fan with no rotation compensation. Only the stationary sound sources from the gap between blades and wall are visible

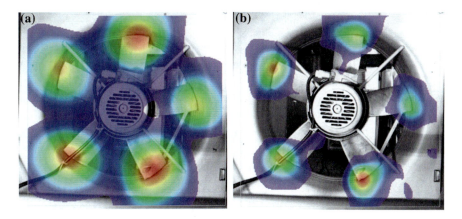

Fig. 7 Analysis with motion compensation. After an alignment of the static optical picture the moving sources at the blade tips are visible. **a** $f = 2{,}000$ Hz. **b** $f = 4{,}000$ Hz

Sum Beamforming algorithm shows a ring shaped distribution of sound sources in the gap between the blades and the wall whereas the rotating beamforming shows the spot shaped sound sources on the blades.

Figure 7 shows similar beamforming results like Fig. 6, but with the rotation compensation. Opposite to the beamforming version without rotation compensation stationary sound sources should be averaged out whereas the rotating sources are visible and, in this case, they belong to the sound emitted by the blades itself. Beside this the acoustical photo is superimposed with a frozen-image to match the sound sources to the blades of the fan.

5 Conclusion

With the here presented rotating beamforming algorithm it is possible to image stationary sound sources as well as rotating sound sources. In combination with an acoustic camera it is a helpful tool for optimising fan geometries to reduce sound emission.

With this algorithm it is possible to locate sound sources at their position on the blades. So it is possible to distinguish between the leading edge and the trailing edge of the blade and study the frequency dependence of the generated noise.

References

1. Christensen, J.J., Hald, J.: Technical review No.1 2004- beamforming. Brüel&kjær sound& vibration measurement A/S. http://www.bksv.com/doc/bv0056.pdf (2004). Accessed 20 Dec 2012

2. Sijtsma, P. Oerlemans, S. Holthusen, H.: Location of rotating sources by phased array measurements. National Aerospace Laboratory (2001)
3. Brooks, T. F., Humphreys, W. M. Jr.: A deconvolution approach for the mapping of acoustic sources (DAMAS) determined from phased microphone arrays. In: 10th AIAA/CEAS aeroacoustics conference (2004)
4. Pannert, W.: Rotating beamforming-motion-compensation in the frequency domain and application of high-resolution beamforming algorithms. Report University of Applied Scienes Aalen. (to be published in Journal of Sound and vibration in 2013)
5. Ehrenfried, K.: Skript zur vorlesung strömungsakustik (script to the teaching lesson fundamentals of aeroacoustics). Mensch and Buch Verlag. http://vento.pi.tu-berlin.de/STROEMUNGSAKUSTIK/SCRIPT/nmain.pdf (2004). Accessed 20 Dec 2012
6. Kenneth, S.B.: Numerical algorithms for acoustic integrals- the devil is in the details. In: 2nd AIAA/CEAS aeroacoustics conference (17th AIAA Aeroacoustics). http://citeseerx.ist.psu.edu/viewdoc/summary?doi=10.1.1.45.6510 (1996). Accessed 20 Dec 2012
7. Brentner, K.S. and Holland, P. C.: An efficient and robust method for computing quadrupole noise. 2nd International aeromechanics specialists. http://www.ingentaconnect.com/content/ahs/jahs/1997/00000042/00000002/art00007 (1995). Accessed 20 Dec 2012
8. Advanced Rotorcraft Technolgy, Inc.: Kirchhoff code—a Versatile CAA Tool. NASA SBIR Phase I Final Report, contract NAS1-20,366 (1995)

Mathematical Modelling of the Physical Phenomena in the Interelectrode Gap of the EDM Process by Means of Cellular Automata and Field Distribution Equations

Andrzej Golabczak, Andrzej Konstantynowicz and Marcin Golabczak

Abstract In this chapter a new attitude has been presented to the mathematical modelling of the physical processes in the interelectrode gap which takes place during electroerosive machining. It relays on use of the partial differential equations describing distribution of fields in the interelectrode gap, covering potential, potential gradient (electric field intensity) and current spread. The samples of the numerical calculations have been presented for the real conditions of the EDM process: dimenstions of the interelectrode gap and roughness of the surfaces of an electrode and machined material. In the analysis some experimental results gathered so far have been used.

Keywords EDM · Mathematical model · Surface roughness · Potential gradient · Numerical calculations

---

A. Golabczak (✉) · A. Konstantynowicz
Lodz University of Technology, Department of Production Engineering,
Stefanowskiego 1/15 Str 90-924 Lodz, Poland
e-mail: andrzej.golabczak@p.lodz.pl

A. Konstantynowicz
e-mail: andrzej.konst@gmail.com

M. Golabczak
Lodz University of Technology, Institute of Machine Tools and Production Engineering,
Stefanowskiego 1/15 Str 90-924 Lodz, Poland
e-mail: marcin.golabczak@p.lodz.pl

1 Introduction

The electro discharge machining (EDM) is particularly well fitted for the precise shaping of machine elements, especially made of very hard (hardly machinable) construction materials conducting electrical current. In the EDM the voltage difference in the form of the train of regulated impulses is applied between the electrode (erode) and machined material. In this process the disposal of the machining allowance is forced by the phenomena correlated with the impulse electric discharge in the interelectrode gap, comprised, among others, of melting, evaporation and abduction of the tiny particles of machined material as a result of imposed strains. The interelectrode gap is filled with the dielectric liquid which maintains: transportation of the eroded material out of the gap as well as cooling of the machined surface and the electrode, and proper spatial shaping of the discharge between the electrode and machined material.

Physical conditions in which electroerosion is conducted (takes place) have been exposed in details both from theoretical and experimental point of view in many works, e.g. [1, 4, 7, 12]. The authors of this chapter carefully investigated electrical phenomena of the EDM, especially related to the origin of the discharge. For the selection of the proper parameters of the EDM, assuring attaining of the prescribed technological tasks (precision of the machining, surface roughness, outer layer state) and economic (productivity, energy consumption), it is necessary to elaborate proper mathematical models of the phenomena occurring during this process. Undoubtedly, there are a number of constituent technological sub processes but the sub process of discharge between electrode and machined material is at the head, especially its initialization. The subsequent stages are: formation of the cathode spot and very complex phenomena companying destruction of the material in this location, i.e. melting, evaporating, sublimation, interaction among charged products of decomposition with electromagnetic field, and finally—their diffusive spread in the dielectric liquid. The final stages of a unique „electroerosion act" are cooling and stabilizing of the machined material surface and sweeping out erosion products from the gap [7].

2 Mathematical Modelling of the Fields in the Interelectrode Gap

From the analysis of the physical foundations of the electrical phenomena taking place in the interelectrode gap during EDM, it could be carried out, that for the modelling of the discharge arising in the interelectrode gap, the proper and suitable tools would be [5, 6, 9]:at the pre-discharge state (without electric current in the gap)—Laplace equation:

Mathematical Modelling of the Physical Phenomena

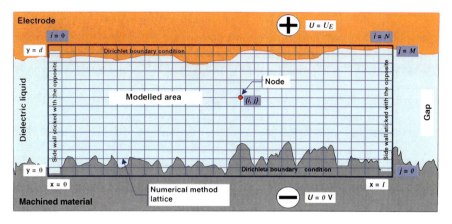

Fig. 1 Schema of the modelled environment—cross-section through the gap and boundaries of the electrode and material

$$\frac{\partial^2 u(x,y,z)}{\partial x^2} + \frac{\partial^2 u(x,y,z)}{\partial y^2} + \frac{\partial^2 u(x,y,z)}{\partial z^2} = 0 \qquad (1)$$

at the discharge initiation stage, during formation of the spatial charge (both from electron cold emission from the cathode and from possible ionization of the dielectric particles)—Poisson equation:

$$\frac{\partial^2 u(x,y,z)}{\partial x^2} + \frac{\partial^2 u(x,y,z)}{\partial y^2} + \frac{\partial^2 u(x,y,z)}{\partial z^2} = -\frac{\rho(x,y,z)}{\varepsilon_0 \cdot \varepsilon} \qquad (2)$$

where:

$u(x, y, z)$ potential function of the electrostatic field in the gap, in V,
$\rho(x, y, z)$ charge spatial distribution (charge density) function in the gap, in C/m^3,
ε_0 vacuum dielectric permittivity, in F/m, $\varepsilon_0 = 8.85418782 \cdot 10^{-12}$ F/m,
ε relative dielectric permittivity of the dielectric liquid, (for the distilled water at 293 K, $\varepsilon \approx 80$)

The partial differential equations mentioned above are a particular case of the Maxwell[1] equations which could be used for the modelling of the EDM phenomena in the gap [5, 9]. This can be justifed by the distance scale at which the modelled phenomena arises (gap thickness in EDM ranges from 30 µm to 500 µm) and the fact, that velocities of all the particles are in the non-relativistic range. It means that we are out of the area reserved for quantum mechanics. The spatial structure of the modelled fragment of the EDM gap has been depicted on Fig. 1.

---

[1] Maxwell (J. Clerk) published his equations in the article "*A Dynamical Theory of the Electromagnetic Field*" in 1864.

The first attempt has been restricted to the two-dimensional area in the gap cross-section. This does not diminish the modelling accuracy and generality, but makes easier a pictorial form of presentation. At the beginning stage of machining the spatial distribution of discharges on the machined surface is of random type [7, 12], which points its strong relation with the geometrical structure of the surface. The surface roughness is then a Dirichlet boundary condition for the Eqs. (1) and (2), i.e., on the whole electrode surface we assume the same positive potential and on the whole machined surface we assume zero potential.

$$u(x,y)|_{y=0} = 0 \qquad u(x,y)|_{y=d} = U_E \qquad (3)$$

The assumption that the conductivity of the electrode material as well as the machined material is of a few magnitude levels higher than conductivity of the dielectric liquid, gives an additional strong justification for the Dirichlet boundary condition of the form (3).

Establishing the boundary conditions for the side walls of the modelled area is not so unambiguous. For sure it is not the Dirichlet boundary condition. It could be assumed that this is a Neumann condition, i.e. a particular value of the spatial derivative of the potential function, equal with the average field gradient in the gap:

$$\left.\frac{\partial u(x,y)}{\partial y}\right|_{x=0} = \left.\frac{\partial u(x,y)}{\partial y}\right|_{x=l} = \frac{U_E}{d} \qquad (4)$$

But still the influence of such a boundary condition on the disturbance of the potential distribution inside the gap is disputable. The problem of the side boundary condition can be solved also in a different way. "Sticking" together the left and right wall of the modelled area, one impose on the modelled area an "infinity" feature and the boundary problem for the side wall simply does not matter.

Solving of Eq. (1) allows one to describe the place in which the electrical discharge will arise—this will be the point with the greatest value of the potential gradient $\mathbf{E}(x, y)$, i.e. this point at which function (5) will take the greatest value:

$$\mathbf{E}(x,y) = -\operatorname{grad} u(x,y) = -\left(\vec{i} \cdot \frac{\partial u(x,y)}{\partial x} + \vec{j} \frac{\partial u(x,y)}{\partial y}\right) \qquad (5)$$

In electrostatic fields (i.e. without curls) it is assumed that in every time moment t and at every point of space (x,y), the current density $\mathbf{J}(x, y, t)$ in stationary state, when the current is practically invariable, is related to the environment conductivity $\sigma(x, y)$ through the relationship (6):

$$\mathbf{J}(x,y,\tau) = \sigma(x,y) \cdot \mathbf{E}(x,y,\tau)|_{\tau=t} \qquad (6)$$

Unfortunately, in our case, relation (6) could not be applied because of:

the diffusion of the charge carriers in the dielectric, is forced also by the gradient of its' density, not only by an abduction in the electric field, the backward interaction of the charges distribution on the potential field distribution is described Eq. (2).

Fig. 2 An example of a discharge in dielectric solid, so-called Lichtenberg's figure

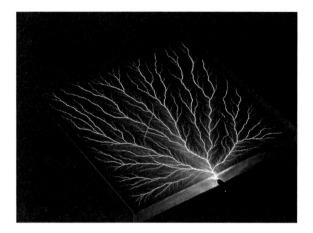

We have to start from the more fundamental dependence:

$$\mathbf{J}(x, y, \tau) = q \cdot n(x, y, \tau) \cdot \mathbf{V_D}(x, y, \tau)|_{\tau=t} \tag{7}$$

where:
- q elementary charge (of electron): $q = 1.60217733 \cdot 10^{-19}$ A·s,
- $n(x, y, \tau)$ charge distribution, in $[1/m^3]$,
- $\mathbf{v_D}(x, y, \tau)$ distribution of the entrainment (drift) velocity of the charged particles in the electric field of the intensity $\mathbf{E}$, and diffusion forced by the gradient of density (concentration), in [m/s]

The correct formulation of the currents' flow at the moment of discharge origination (2nd stage), i.e. settlement of the dependencies between $\mathbf{v_D}$ and $\mathbf{E}$, seems to be very difficult and will be the subject of our further efforts in modelling. The scale of difficulty could be evidenced in Fig. 2, which depicts the so-called Lichtenberg's figure arising when the discharge is conducted in the translucent solid of the dielectric. It is named after the German physicist who first was able to "freeze" the view of electric discharge.

The generation of such a figure by using the mathematical model is a serious task itself, so more that the presented figure does not exhibit a diffusion of the charge carriers in time—the discharge has been "frozen" in the acrylic resin. One can see only a result of the dynamical process, not a line of development of structure in time.

2.1 Numerical Solution

For the solution of Eqs. (1) and (2), numerical methods have been chosen. Analytical methods are not appropriate, even as a theoretical alternative, because the boundary problem described by the expression (3) has a strongly random character and modelling of this curve is a serious task itself, relatively complex, which we will deal with on the further pages.

The relaxation method has been applied [8] on a rectangular lattice with dimensions N × M nodes, placed upon the area of real dimensions $l \times d$ (as was depicted on Fig. 1). The numerical solution was implemented (programmed) in a *Excel* calculation sheet. The main and unquestioned profit of such an strategy is that the solution is numerically stable a unique (unambiguous), i.e. what we can see is really a solution. Describing in details, it is based on:

- substituting of the differential equation with the equivalent difference equation which approximates it in one node of lattice,
- solving of the linear equations' set for all the nodes in the lattice.

For each node numbered (i, j), the derivatives in Eq. (1) are approximated by the appropriate difference quotient:

$$\frac{\partial^2 u(x,y)}{\partial x^2} \approx \frac{u(x_{i-1}, y_j) - 2.u(x_i, y_j) + u(x_{i+1}, y_j)}{\Delta x^2}$$
$$\frac{\partial^2 u(x,y)}{\partial y^2} \approx \frac{u(x_i, y_{j-1}) - 2.u(x_i, y_j) + u(x_i, y_{j+1})}{\Delta y^2} \quad (8)$$

where:
Δx The lattice step in the direction X, i.e. $\Delta x = l/N$,
Δy The lattice step in the direction Y, i.e. $\Delta y = d/M$,

Summing of the difference quotients (8) in the node (i, j) one obtains the difference equation in this node:

$$\frac{u(x_{i-1}, y_j) - 2.u(x_i, y_j) + u(x_{i+1}, y_j)}{\Delta x^2} + \frac{u(x_i, y_{j-1}) - 2.u(x_i, y_j) + u(x_i, y_{j+1})}{\Delta y^2} = 0 \quad (9)$$

Proper ordering of (9) yields:

$$[u(x_{i-1}, y_j) + u(x_{i+1}, y_j)] \cdot \frac{\Delta y^2}{2 \cdot (\Delta x^2 + \Delta y^2)} + [u(x_i, y_{j-1}) + u(x_i, y_{j+1})]$$
$$\cdot \frac{\Delta y^2}{2 \cdot (\Delta x^2 + \Delta y^2)} = u(x_i, y_i) \quad (10)$$

For the symmetric, square network, Eq. (10) takes the well known, simple form:

$$\frac{u(x_{i-1}, y_j) + u(x_{i+1}, y_j) + u(x_i, y_{j-1}) + u(x_i, y_{j+1})}{4} = u(x_i, y_i) \quad (11)$$

For the numerical calculations a rectangular lattice of 150 × 100 nodes has been applied, containing 15,000 linear equations of the type (11), solved

Mathematical Modelling of the Physical Phenomena 175

iteratively. It takes about 3 min in average of the computer work to attain the relative accuracy of the solution at the level of 10^{-3}. A rectangular lattice, not square, has been applied to allow further pliant scaling of the electrode gap area as depicted in Fig. 1.

The results of the calculations are shown in the subsequent figures: Fig. 3—distribution of the electric potential in the gap, Fig. 4—potential gradient (electric field intensity) at the machined surface side, Fig. 5—potential gradient (electric field intensity) at the electrode surface side.

In Fig. 4 there are a few of very characteristic ranges placed between edges of roughness. Inside them a very low electric potential gradient is evidenced, considerably lower than the average electric field strength in the gap—this is the so-called "shielding effect". It exhibits clearly, that the EDM method has a natural tendency for leveling of the surface unevenness, not for deepen them.

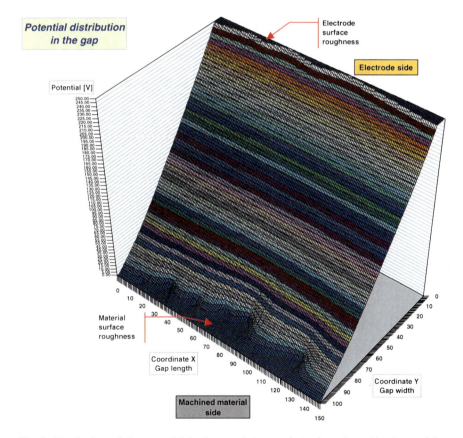

Fig. 3 Distribution of the potential in the gap between electrode and machined material at conditions: width of the gap 40 µm, voltage at the electrode +250 V

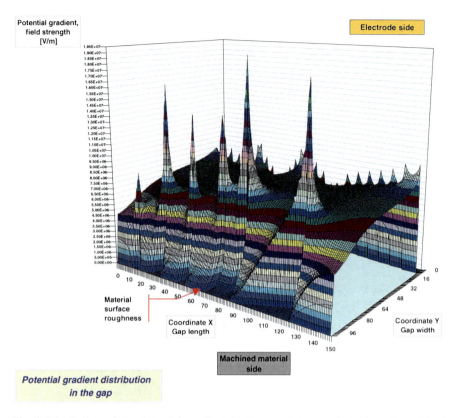

Fig. 4 Distribution of the potential gradient in the gap between electrode and machined material—view from the material side; conditions as previous on Fig. 3

3 Modelling of the Surface Roughness

The way of surface roughness modelling we propose in this chapter is a relatively modern, based on the so-called "cellular automata" approach, also called the "mosaic automata". The inventor of this skillful idea of dynamical systems description was S. M. Ulam.[2] Cellular automata are completely discrete dynamical systems. Their "discreetness" is both of spatial and time nature. They operated on the surface tilled into rectangular (in our case) cells which state the change only in discrete time moments and usually is expressed by integer numbers, although the latest condition is not essential for the automaton behavior. What is settled by the automaton designer are so-called transition rules, i.e. the rule of change of the

---

[2] Stanisław Marcin Ulam (1909–1984) magnificent polish mathematician from "Lvov school", also the author of very advanced works in the number theory as well as other numerical methods of the so—called Monte Carlo type. The majority of his works have been done at Los Alamos where he was occupied with the project of thermonuclear bomb development.

Mathematical Modelling of the Physical Phenomena

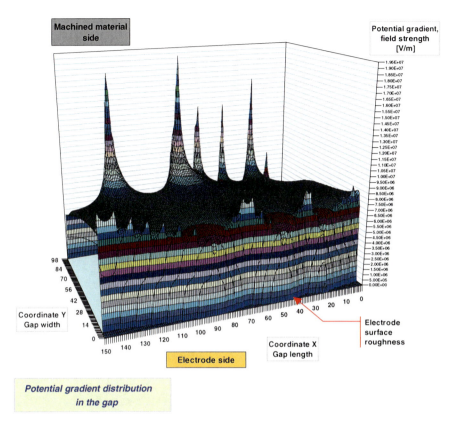

Fig. 5 Distribution of the potential gradient in the gap between electrode and machined material—view from the electrode side; conditions as previous on fig. 3

given cell state accordingly to its previous state and the states of surrounding cells in the closest vicinity. To this category of dynamical systems belongs also an intellectual toy called "Game of Life" devised by John Horton Conway in 1970 as a two-dimensional automaton used by biologists for the modelling of the population dynamics of living creatures [3].

Our subject of interest in the area of surface modelling is the so-called one-dimensional automata—extremely simple but very fruitful in their activity. The one of possible transition rules has been illustrated in Fig. 6. Total number of rules for a given automaton is surprisingly big:

$$(number\ of\ states)^{(number\ of\ states)^{(number\ of\ previous\ neighbours)}} = 2^{2^3} = 256 \qquad (12)$$

These states are numbered accordingly with the natural binary code and hence rules used by different authors, called rule "30" or rule "110", are adequate for binary numbers consisting of 3 bits:

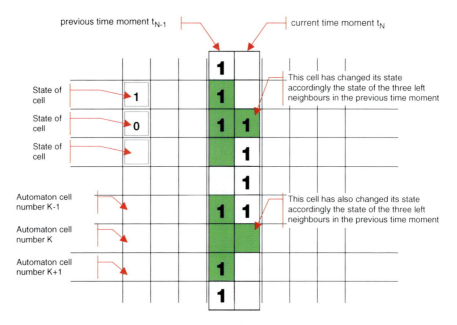

Fig. 6 Rules of the construction of the one-dimensional cellular automaton; values of states have been exhibited for the rule "30"

$$30_{dec} = 00011110_{bin}$$
$$110_{dec} = 01101110_{bin} \qquad (13)$$

Units in the binary expansion of the rule number describe for which of the left neighbors combinations the next state of the given cell will be equal to one.

In Fig. 7, an example has been exhibited of the roughness modelling by using of the cellular one-dimensional automaton acting accordingly to the rule "30" with random initial state. Although this result looks very "probable" the important question arises if this attitude to the surface modelling is adequate to the specific roughness which arises during EDM? The answer demands thorough experimental verification which will be the subject of our further works. Nevertheless, at this level of our knowledge, it should be stressed, that application of a cellular automaton as a "machine" for surface pattern generation gives the pattern which is more repeatable than any other numerical process, thus its comparability seems to be advantageous.

4 Mathematical Modelling of the Discharge Development

The mathematical modelling of the discharge development in the gap demands the formulation of the two main sub problems:

Fig. 7 Surface roughness pattern generated with one-dimensional cellular automaton at random initial state and rule "30" applied for system transition

- transportation of the processed material through the gap, in the form of irregular grits rended from the processed material, by the diffusion process as well as the abduction in the electric field,
- electric field distribution with taking into account spreading of the grits with their dielectric permittivity and possible charging.

The modelling of the grits movement through the gap can be done with relatively great accuracy by using proper cellular automaton of the probabilistic type which allows to accomplish the classical diffusion modelling as well as the field imposed abduction [2, 10]. We used one-layer automaton with Moore vicinity of cells, as depicted in Fig. 8. The automaton grid is the same as for the electric field distribution numerical solving, depicted in Fig. 1. The basic movement of one grit per one step of simulation is 1 cell aside. This is not a so strong constraint as one could imagine, because this can be simply taken into account when establishing the time scale of automaton, as described later.

All of the transition probabilities $P(i, j)$ are constituted of two parts: the first $P_0(i, j)$ is responsible for the diffusion mobility and the second $P_E(i, j)$ is responsible for the field abduction, being the function of field strength (potential gradient) $E(i, j)$, grit charge density $\sigma(i, j)$ and grit dipole momentum $\pi(i, j)$ induced by the electric field:

$$P(i,j) = P_0(i,j) + P_E(i,j) \qquad (14)$$

$$P_E(i,j) = f(E(i,j), \sigma(i,j), \pi(i,j)) \qquad (15)$$

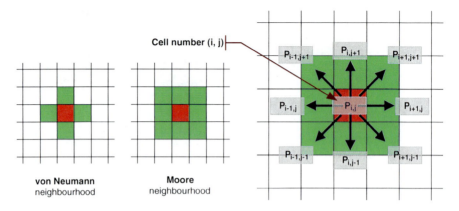

Fig. 8 Basics of the probabilistic cellular automaton used for modelling grits movement by means of diffusion as well as the electric field abduction. Transition probabilities have been defined for one of the cells

The first, "permanent" part of Eq. (14) is responsible for the time scale of the process model and the Brownian-like behavior of the grits. When the "self-transition" $P_{i,j}$ is relatively big in comparison with transitions $P_{i\pm 1, j\pm 1}$, the Brownian motion is slow, i.e. the grit RMS displacement per one time step is small. Basing on the proportion: $\frac{P_{i,j}}{\sum P_{i\pm 1, j\pm 1}}$ one can set movement rules for the electric abduction, i.e. to set up a function in Eq. (15) maintaining the force-induced behavior of the grit. Considering the electric field distribution calculated for the global distribution of all of the grits we can also model the grits' interaction in this way.

One more condition has to be imposed on the set of probabilities in Eq. (14), for every cell (i, j), which is obvious if the automaton has to be probabilistic:

$$\sum \left(P_{i,j} + P_{i\pm 1, j\pm 1} \right) = 1 \tag{16}$$

For solving the electric field distribution we use the more basic form of the Poisson Eq. (2) allowing the consideration of the dielectric permittivity of all components of the physical process as well as its possible charging, as follows:

$$\mathrm{div}(D(x, y)) = -\sigma(x, y) \tag{17}$$

where:
$D(x, y)$ Dielectric displacement is given as follows:

$$(D(x,y)) = \varepsilon(x,y) \cdot E(x,y) = \varepsilon(x,y) \cdot \mathrm{grad}(u(x,y)) \tag{18}$$

where $\varepsilon(x, y)$ is generally a tensor but in our case we perform the modelling in an isotropic liquid and then we assume that it is a scalar spatial function, which in turn leads to the following development of the expression (18)

$$\mathrm{div}(\varepsilon(x,y) \cdot \mathrm{grad}(U(x,y))) = \varepsilon(x,y) \cdot \mathrm{div}(\mathrm{grad}(U(x,y))) + \mathrm{grad}(u(x,y)) \\ \cdot \mathrm{grad}(u(x,y)) \tag{19}$$

After substituting expression (19) into (17) one can obtain in the full form:

$$\varepsilon(x,y) \cdot \left(\frac{\partial^2 U(x,y)}{\partial x^2} + \frac{\partial^2 U(x,y)}{\partial y^2} \right) + \frac{\partial U(x,y)}{\partial x} \cdot \frac{\partial \varepsilon(x,y)}{\partial x} + \frac{\partial U(x,y)}{\partial y} \cdot \frac{\partial \varepsilon(x,y)}{\partial y} \\ = -\sigma(x,y) \tag{20}$$

Equation (20) can be transformed into its numerical counterpart, accordingly rules presented for Eq. (10), which yields:

$$U(x_i, y_i) = \\ \frac{\Delta y^2}{\Delta x^2 + \Delta y^2} \cdot \left(\frac{1}{2} + \frac{\varepsilon(x_{i+1}, y_j) - \varepsilon(x_{i-1}, y_j)}{8 \cdot \varepsilon(x_i, y_j)} \right) \cdot U(x_{i+1}, y_j) + \\ \frac{\Delta y^2}{\Delta x^2 + \Delta y^2} \cdot \left(\frac{1}{2} - \frac{\varepsilon(x_{i+1}, y_j) - \varepsilon(x_{i-1}, y_j)}{8 \cdot \varepsilon(x_i, y_j)} \right) \cdot U(x_{i-1}, y_j) + \\ \frac{\Delta y^2}{\Delta x^2 + \Delta y^2} \cdot \left(\frac{1}{2} + \frac{\varepsilon(x_i, y_{j+1}) - \varepsilon(x_i, y_{j-1})}{8 \cdot \varepsilon(x_i, y_j)} \right) \cdot U(x_i, y_{j+1}) + \\ \frac{\Delta y^2}{\Delta x^2 + \Delta y^2} \cdot \left(\frac{1}{2} - \frac{\varepsilon(x_i, y_{j+1}) - \varepsilon(x_i, y_{j-1})}{8 \cdot \varepsilon(x_i, y_j)} \right) \cdot U(x_i, y_{j-1}) - \frac{1}{8 \cdot \varepsilon(x_i, y_j)} \frac{\Delta x^2 \cdot \Delta y^2}{\Delta x^2 + \Delta y^2} \cdot \sigma(x_i, y_j) \tag{21}$$

Solving of Eq. (21) has been performed by using the same relaxation method as previously. The numerical results are been presented in Figs. 9 and 10. For the simulation process we assume corundum as the processed material and kerosene as the working fluid in the gap, with dielectric permittivities $\varepsilon = 10.5$ and $\varepsilon = 1.8$, respectively. The electrode material was copper as in previous simulations. The real dimension of the lattice cell was 0.4 µm, which seems to allow correctly reflect the natural conditions of the EDM process.

The decision on which part of material to rend at every stage of simulation we performed on a probabilistic base, assuming that the probability of rending rises strongly when the field strength is greater than 10^6 V/m and becomes "1" in the field 10^7 V/m. The dimension of the rended grit we assumed (at this stage of our survey) to be 1 or 2 cells.

Every stage of simulation consisted of three steps:

- calculation of the electric field distribution in the gap,
- decision which part of material will be rend by the electric field,
- random drawing of grits' movement, 1 cell aside.

The distribution of the potential of the electric field, depicted in Fig. 9. is slightly enlarged at the processed material side to make the grits-induced disturbances more visible. Careful observation of the potential distribution in the grit

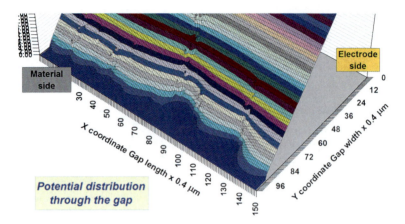

Fig. 9 Potential distribution in the gap during discharge development. Field disturbances caused by charged grits with dielectric permittivity higher then fluid in the gap are clearly visible

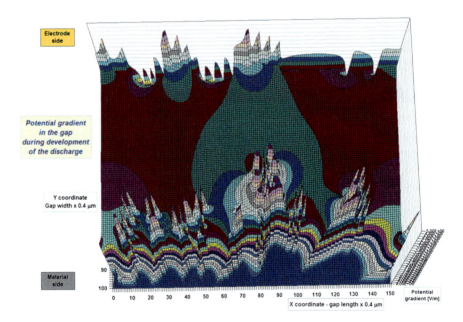

Fig. 10 Potential gradient in the gap during development of the discharge. The discharge channel begin to form

nearest vicinity leads to the conclusion that field-induced charging depends upon the ratio of the dielectric permittivities of the material and fluid, plays even a more important role in the grit movement than the electrical charge carried from the material. This in turn, leads to the preliminary conclusions due to the electrode polarization accordingly mutual dielectric permittivities of the material and fluid in the processing gap.

5 Summary

The method (or even methodic) of the mathematical modelling of the electroerosion process in the electrode gap during EDM was presented. The numerical evaluations have been carried out for the real, technological conditions The introductory results related to the potential distribution, the field strength distribution, the current flow in the electrode gap and the surface roughness are promising enough to justify the need for further works, directed especially for the experimental verification of the presented modelling operational usefulness. The very first observations have lead us to the following preliminary conclusions. Cross-use of the cellular automaton and Poisson equation solved on the same grid has given us a very efficient tool for modelling discharge development, including:

- material rending from processed material,
- grits propagation through dielectric liquid,
- interactions among grits,
- spatial distribution of the electric field as well as it's time changes.

References

1. Albinski, K., Bratasz, Ł. et al.: Promieniowanie plazmowego kanału wyładowania elektrycznego. In: International Conference EM'97, Bydgoszcz, Poland (1997)
2. Chassaing, P., Gerin, L.:Asynchronous cellular automata and Brownian motion. In: Conference on Analysis of Algorithms A of A 07 (2007)
3. Gobro, S., Chiba, N.: Crack Pattern simulation based on 3D surface cellular automata. The Visual Computer 17(5), 287–309 (2001)
4. Gołąbczak, A., Kozak, J.: Studies of Electrodischarge and electrochemical systems for dressing of metal bond of grinding wheels. J. Eng. Manuf. (2006)
5. Konorski, B.: Elementy teorii względności, relatywistycznej mechaniki i elektrodynamiki dla inżynierów. WNT, Warszawa (1976)
6. Konorski, B.: Podstawy elektrotechniki. PWN, Warszawa (1967)
7. Miernikiewicz, A.: Doświadczalno-teoretyczne podstawy obróbki elektroerozyjnej EDM. Kraków, Poland (2000)
8. Southwell, R.V.: Relaxation methods in theoretical physics. Clarendon Press, Oxford (1956)
9. Vanderbilt, D.: Polarization, electric fields and dielectric response in insulators. Conference on Computational Physics, Rutgers University (2005)
10. Zaloj, V., Agmon, M.: Electrostatics by Brownian: solving the Poisson equation near dielectric interfaces. Chem. Phys. Lett. 270, p. 476 (1997)

11. Zhenli, X.: Electrostatic interaction in the presence of dielectric interfaces and polarization-induced like-charge attraction. Cornell University Library, NewYork (2013)
12. Zolotych, B.N.: Osnovnyje voprosy kacestviennoj teorii elektroiskrovoj obrabotki v židkoj dielektriceckoj srede. In: Problemy elektriceskoj obrabotki metallov, Moscov (1962)

Free Vibration Analysis of Clamped-Free Composite Elliptical Shell with a Plate Supported by Two Aluminum Bars

Levent Kocer, Ismail Demirci and Mehmet Yetmez

Abstract In this chapter, free vibration analysis of a clamped-free E-glass composite elliptical shell with an interior carbon composite plate is studied. At one end of the carbon/epoxy plain weave composite plate, two Al2024-T3 bars are attached. Vibration tests are performed to present the free vibration characteristics of the clamped-free composite structure provided by TAI-Turkish Aerospace Industry Inc. Effects of the structural parts on the vibration characteristics are examined both experimentally and numerically. In order to explore the possibilities for further finite element research on such vibration analysis, the sub-structural effects on the numerical model is considered in a limited manner.

Keywords Elliptical shell · Plain weave · Frequency response · Finite elements

1 Introduction

When any composite material is subjected to dynamic forces, it vibrates. Very often the vibrations have to be investigated, either because they cause an immediate problem, or because the structure has to be cleared to a standard or test specification.

L. Kocer · I. Demirci
Eregli Iron and Steel Works Co., Zonguldak, Turkey
e-mail: lkocer@erdemir.com.tr

I. Demirci
e-mail: ismaildemirci@erdemir.com.tr

M. Yetmez (✉)
Zonguldak Karaelmas University, Zonguldak, Turkey
e-mail: mehmet.yetmez@karaelmas.edu.tr

By using signal-analysis techniques, vibration can be measured on the operating structure and make a frequency analysis. The frequency spectrum description of how the vibration level varies with frequency can then be checked against any specification. This type of testing gives results which are only relevant to the measured conditions. The result obtained is a product of the structural response and the spectrum of an unknown excitation force.

An approach is generally known as a system analysis technique where a dual-channel Fast Fourier Transform analyzer is used to measure the ratio of the response to a measured input-force. Then once vibration testing data including excitation and response signals in time domain are collected, the frequency response function (FRF) can be easily computed. It is recollected that the FRF measurement removes the force spectrum from the data and describes the inherent structural response between the measured points [1]. Additionally, the vibration problems of composite shell and plate are studied using various methods such as the receptance method (i.e., FRF) [2–5]. Furthermore, one can summarize that the current experimental achievements in both shell and plate structures are limited for the sake of damage detection [6, 7].

For the further damage-based studies, the purpose of this study is to examine effects of the structural parts of an E-glass composite elliptical shell without any damage both experimentally and numerically. Additionally, in order to explore the possibilities for further finite element research on such vibration analysis, the substructural effects on the numerical model is considered in a limited manner.

2 Experimental Procedure

The shell structure consists of three materials: an E-glass composite elliptical shell, a plain weave composite plate and an aluminum bar. The mechanical properties of the materials of the shell structure are presented in Table 1. From the vibration testing point of view, first and second natural frequencies of the clamped-free E-glass composite elliptical shell with an interior carbon composite plate is measured by the following procedure.

An impact hammer with a force transducer (Model No: 5800B2, Dytran Instruments, Inc., USA) is used to excite the E-glass composite elliptical shell with an interior carbon composite plate and two aluminum bars given in Fig. 1. Two impact excitations are applied to the selected point of the shell as shown in Fig. 2.

Table 1 Mechanical properties of the components of the shell structure

Material type	ρ (kg/m^3)	E_{yy} (GPa)	υ_{xy}
E-glass elliptical shell	2000	43.5	0.27
Carbon plain weave plate	1671	49.691	0.28
Al2024-T3 aluminum bar	2770	71	0.33

Free Vibration Analysis of Clamped-Free Composite Elliptical Shell 187

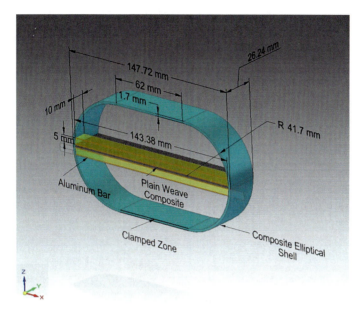

Fig. 1 Geometrical representation of a clamped-free E-glass composite elliptical shell with an interior carbon composite plate and two aluminum bars

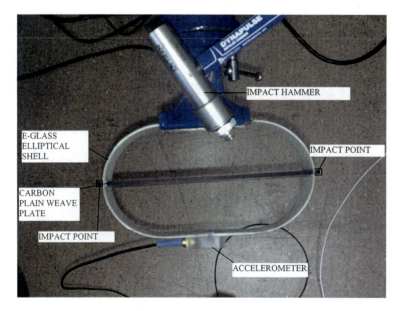

Fig. 2 Test set-up of the clamped-free E-glass composite elliptical shell with an interior carbon composite plate

After the excitations, the responses are obtained by an accelerometer (Model No: 3093B, Dytran Instruments, Inc., USA).

The vibration measurements are completed using a microprocessor-based data acquisition system, namely SoMat™ eDAQ-lite and nCode GlyphWorks software (HBM, Inc., USA).

The accelerometer is located at the middle of the free side of the shell in all measurements. An impact force of 15 N is applied to the middle point between the clamped region and the accelerometer location for all test specimens and all impact excitations.

3 Numerical Solution

The general purpose finite element code ANSYS is used for the numerical vibration analyses of the free E-glass composite elliptical shell with an interior carbon composite plate and two aluminum bars. It is generally known that the most critical part of these analyses is the finite element model (FEM) which consists of representative volume elements (RVEs). In other words, the RVE plays an important role in the mechanics of composite materials, especially for woven type composites. In this study, a smaller and effective RVE is considered as shown in Fig. 3. Therefore, in order to create a finite element model using both a higher order three-dimensional 20-node solid element (SOLID 186) and a three-dimensional 10-node tetrahedral structural solid element (SOLID187), the proposed

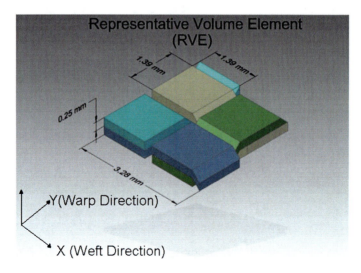

Fig. 3 Representative volume element for the finite element model of the elliptical shell structure

Free Vibration Analysis of Clamped-Free Composite Elliptical Shell

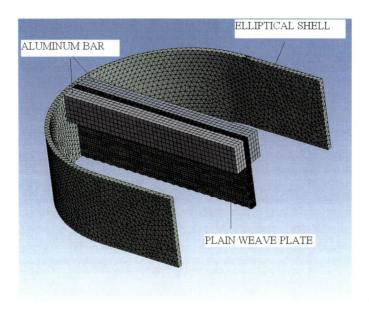

Fig. 4 General representation of the FEM

RVE is taken into account. The general representation of the FEM is given in Fig. 4.

Then a vibration analysis is able to be done with respect to the global stiffness matrix [K] and global mass matrix [M]. The first natural frequencies of the composite plates are calculated by the following equation:

$$[K]\{U\} = \omega^2 [M]\{U\} \qquad (1)$$

In Eq. (1), ω is the natural frequency of the plate considered and $\{U\}$ is the normalized eigenvector. The eigenvalues and eigenvectors are computed by using a preconditioned conjugate gradient Lanczos solver. Also, symmetrical conditions and mass effect of the accelerometer at the free end are to be taken into consideration (see Fig. 4). Whereas the FEM of the E-glass elliptical shell is created by SOLID 187 (# of nodes = 125309, # of element = 69814), that of the E-glass elliptical shell with a carbon plate is created by both SOLID186 and SOLID187 (# of nodes = 542043, # of elements = 63777). Similarly, the FEM of the E-glass

Table 2 First natural frequencies (ω_1) of the FEM of the clamped-free E-glass composite elliptical shell

Structure type	ω_1 (Hz)
E-glass elliptical shell	83.075
E-glass elliptical shell with a carbon plate	111.728
E-glass elliptical shell with a carbon plate and two aluminum bars	102.201

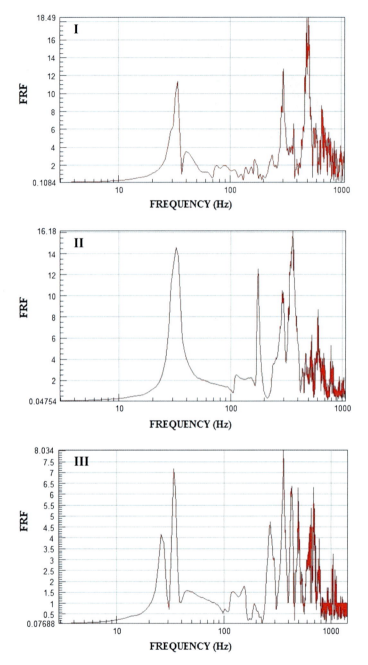

Fig. 5 Frequency response analysis of E-glass elliptical shell (**I**), E-glass elliptical shell with a carbon plate (**II**) and E-glass elliptical shell with a carbon plate and two aluminum bars (**III**) in x-direction

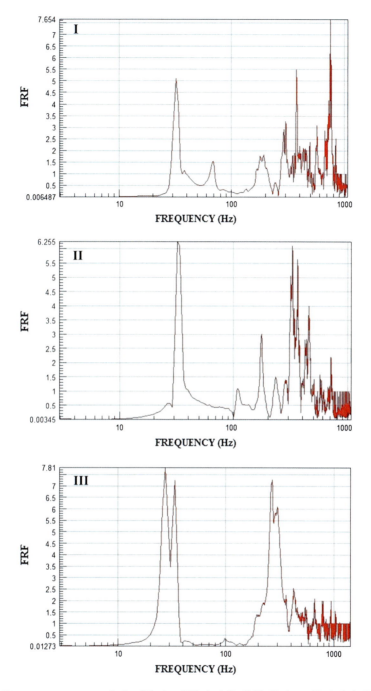

Fig. 6 Frequency response analysis of E-glass Elliptical Shell (**I**), E-glass elliptical shell with a carbon plate (**II**) and E-glass elliptical shell with a carbon plate and two aluminum bars (**III**) in y-direction

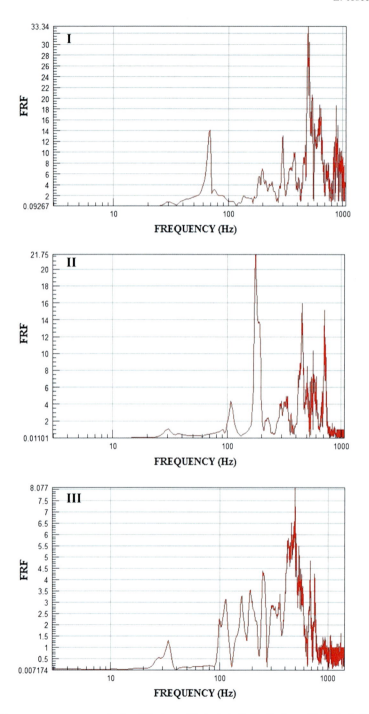

Fig. 7 Frequency response analysis of E-glass elliptical shell (**I**), E-glass elliptical shell with a carbon plate (**II**) and E-glass elliptical shell with a carbon plate and two aluminum bars (**III**) in z-direction

Free Vibration Analysis of Clamped-Free Composite Elliptical Shell

Table 3 Experimental first and second natural frequencies (ω_1 and ω_2) of the clamped-free E-glass composite elliptical shell with an interior carbon composite plate

Structure type	ω_1 (Hz)	ω_2 (Hz)
E-glass elliptical shell	80.92	802.745
E-glass elliptical shell with a carbon plate	128.55	533.385
E-glass elliptical shell with a carbon plate and two aluminum bars	127.476	682.129

Table 4 Comparison of first natural frequencies of the elliptical shell structure between experimental (ω_1^{EXP}) and the finite element (ω_1^{FEM}) results

Structure type	ω_1^{EXP} (Hz)	ω_1^{FEM} (Hz)
E-glass elliptical shell	80.92	83.075
E-glass elliptical shell with a carbon plate	128.55	111.728
E-glass elliptical shell with a carbon plate and two aluminum bars	127.476	102.201

elliptical shell with a carbon plate and two aluminum bars covers not only SOLID186 but also SOLID187 (# of nodes = 553237, # of elements = 65841).

According to the FEM results, first natural frequencies of the clamped-free E-glass composite elliptical shell with an interior carbon composite plate and two aluminum bars are presented in Table 2.

4 Results and Conclusion

Experimental results of the clamped-free E-glass composite elliptical shell with an interior carbon composite plate are presented in Figs. 5, 6, 7.

Through the Figs. 5, 6, 7, it can be noted that the first natural frequencies of structure types I, II and III are approximately 30, 30 and 35 Hz in x-direction; 32, 35 and 25 Hz in y-direction; 68, 120 and 120 Hz in z-direction, respectively (see Fig. 1). Similarly, it can be easily seen that the second natural frequencies of those types I, II and III are around 500, 360 and 370 Hz in x-direction; 380, 350 and 280 Hz in y-direction; 500, 180 and 500 Hz in z-direction, respectively.

According to the results, two natural frequencies of the clamped-free E-glass composite elliptical shell with an interior carbon composite plate are presented in Table 3.

Table 4 presents the comparison of fundamental frequencies of the elliptical shell structure between experimental (ω_1^{EXP}) and the numerical (ω_1^{FEM}) results. The results indicate that the characteristic behavior of numerical predictions (i.e., FEM) is compatible with that of experimental evaluations especially for the first two structural types, namely E-glass elliptical shell and E-glass elliptical shell with a carbon plate. In other words, FEM for the two structure types seems to work effectively.

In addition to that, although FEM for the last structure type (i.e., E-glass elliptical shell with a carbon plate and two aluminum bars) shows a little bit a similar trend regarding to the experimental evaluations, results of FEM are good for the numerical prediction.

References

1. Dossing, O.: Structural testing, part II: modal analysis and simulation. Brüel & Kjaer, Naerum (1988)
2. Azimi, S., Hamilton, J.F., Soedel, W.: The receptance method applied to the free vibration of continuous rectangular plates. J. Sound. Vib. **93**(1), 9–29 (1984)
3. Huang, D.T., Soedel, W.: Study of the forced vibration of shell-plate combinations using the receptance method. J. Sound. Vib. **166**(2), 341–369 (1993)
4. Yim, J.S., Shon, D.S., Lee, Y.S.: Free vibration of clamped-free circular cylindrical shell with a plate attached at an arbitrary axial position. J. Sound Vib. **213**(1), 75–88 (1998)
5. Lee, Y.S., Choi, M.H.: Free vibration of a composite cylindrical shell with a longitudinal interior rectangular plate. In: Gakkai, N.Z. (ed.) The first Asian Australasian conference on composite materials. Society of Materials Science, Osaka (1998)
6. Zou, Y., Tang, L., Steven, G.P.: Vibration-based model-dependent damage (delamination) identification and health monitoring for composite structures-a review. J. Sound Vib. **230**(2), 357–378 (2000)
7. Hu, H., Wang, J.: Damage detection of a woven fabric composite laminate using a modal strain energy method. Eng. Struct. **31**, 1042–1055 (2009)

Vibration Analysis of Carbon Fiber T-Plates with Different Damage Patterns

Ismail Demirci, Levent Kocer and Mehmet Yetmez

Abstract In this study, dynamic analysis of carbon fiber T-plates with different damage patterns including cracks and impact-damaged region is considered. For this purpose, vibration tests are performed to present the free vibration characteristics of clamped-free carbon fiber-plain weave composite T-plates provided by TAI-Turkish Aerospace Industry Inc. A general purpose finite element code (ANSYS) is used to confirm the experimentally measured natural frequencies. Effects of damage type, size and location on the vibration characteristics are examined both experimentally and numerically. The thickness effect is also investigated.

Keywords Plain weave · Damage · Impact · Natural frequency · Finite element

1 Introduction

There are many researches to detect damage in plain weave structures with respect to vibration measurements [1, 2]. These investigations conclude that the natural frequencies of a component tend to be reduced by damage. Also, many numerical techniques are used to analysis the vibration characteristics of engineering structures. These researches may briefly summarized as follows: Instead of modelling

I. Demirci · L. Kocer
Eregli Iron and Steel Works Co, Zonguldak, Turkey
e-mail: ismaildemirci@erdemir.com.tr

L. Kocer
e-mail: lkocer@erdemir.com.tr

M. Yetmez (✉)
Zonguldak Karaelmas University, Zonguldak, Turkey
e-mail: mehmet.yetmez@karaelmas.edu.tr

the dynamic behaviour of rectangular composite plates by the finite element method, Deobald and Gibson [3] present the Rayleigh–Ritz technique to model the vibrations of rectangular orthotropic plates and say that the experimental natural frequencies do not always closely match the predicted values. Salawu [4] discusses the relationships between frequency changes and structural damage. Chen and Chou [5] solve natural frequency equations and natural modes for orthogonal-woven fabric composites analytically. Kessler [6] focuses on the relationship between various sensors and their ability to detect changes (i.e., damage) in a material/structure's behavior. For this purpose, two-dimensional finite element models are created for comparison with the experimental results. Considering frequency response functions and a finite element beam model, Maia [7] proposes a new application of some well-known mode-shape-change-based method. Kim [8] works on the residual frequency response functions (FRFs) and the natural frequencies of the structural dynamic system reconstruction of deboned honey-comb sandwich beams and of axial-fatigue-damaged laminated beams. Barbero [9] develops a three-dimensional finite element model using ANSYS [10] and compares the predicted values with the experimental values.

In this paper, effects of damage type, size and location on the vibration characteristics are examined both experimentally and numerically. The thickness effect is also investigated.

2 Materials and Methods

2.1 Experimental Solution

To compare the results of experimental modal analysis with those of the numerical ones, Young's modulus of carbon fiber-plain weave composite plates is to be obtained. On the one hand, six coupon samples for tensile tests are taken into account, i.e., first three of them are four-ply laminate and the last three are five-ply laminate. On the other hand, for vibration tests, four carbon fiber-plain weave T-plates are considered.

At the first part, static tensile tests are carried on a screw-driven tensile testing machine powered by ESIT TB 5000-C3 load-cell and ESIT Data Logger v.1.1.6 (ESIT Electronics Ltd. Co., Turkey). The dimensions of the coupon specimens are $70 \times 13.12 \times 2.548$ mm for the four-ply laminate and $70 \times 13.12 \times 3.06$ mm for the five-ply laminate. For the strain measurements in both x- and y-directions, 3 mm foil resistance strain gages (type : FCA-3-11, gage factor: 2.1 ± 0.01 and gage resistance: $350 \pm 1\Omega$) and an adhesive (P-2) are used. All the strain gages and the adhesive are products of TML, Tokyo Sokki Kenkyojo Co. Ltd., Japan. For each of the six specimens, the strain gage is mounted at the center of the specimen. Then, the 4-channel strain measurements are completed using a microprocessor-based data acquisition system, namely SoMat eDAQ-lite and SoMat™ Test Control Environment software (HBM Inc., USA).

At the second part, first natural frequencies of four plain weave T-plates are measured by the following procedure: An impact hammer with a force transducer (Model No: 5800B2, Dytran Instruments, Inc., USA) is used to excite the undamaged plain weave T-plate given in Fig. 1. Three impact excitations are applied to the selected point of each T-plate. After the excitations, the responses are obtained by an accelerometer (Model No: 3093B, Dytran Instruments Inc., USA).

The vibration measurements are completed using a microprocessor-based data acquisition system, namely SoMat™ eDAQ-lite and nCode GlyphWorks software (HBM Inc., USA).

The four T-plate specimens are clamped to a stable table as an inertia block. Two of them are a four-ply laminate with 2.548 mm in thickness and the last two are a five-ply laminate with 3.06 mm in thickness. Two rubber sheets (13.5 mm × 0.6 mm) are glued at both of the two clamped areas. An accelerometer is located at the middle free end of the plate in all measurements. An impact force of 5 N is applied to the middle point between the clamped region and the accelerometer location for all test specimens and all impact excitations. Experimental results of first natural frequencies of the undamaged T-plates ($\omega_1^{measured}$) are presented in Table 1.

Additionally, corresponding to the specially orthotropic laminate, Eq. (1) is considered to compute the first natural frequency [11];

$$\omega_2^1 = \frac{t^3 \pi^4}{12 \rho} \left[\frac{Q_{11}}{a^4} + \frac{2}{a^2 b^2}(Q_{12} + Q_{66}) + \frac{Q_{22}}{b^4} \right] \quad (1)$$

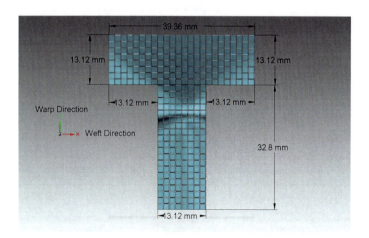

Fig. 1 Geometrical representation of an undamaged plain weave T-plate

Table 1 First natural frequencies of undamaged composite T-plates

Specimen type	ρ (kg/m^3)	E_{yy} (GPa)	υ_{xy}	$\omega_1^{measured}$ (Hz)	$\omega_1^{computed}$ (Hz)
Four-ply	1687	45.912	0.24	347	345.33
Five-ply	1671	49.691	0.28	361	365.028

where ρ, a, b, $Q_{11}, Q_{12}, Q_{22}, Q_{66}$ and ω_1 are the density of the T-plate, width of the plate, length of the plate, reduced stiffnesses of the plate and first natural frequency respectively. Due to the T-plate geometry and mass effect on the fundamental natural frequencies, one may assume that Eq. (1) is reduced to be:

$$\omega_1 = \frac{0.319\pi^2}{b^2}\sqrt{\frac{Q_{22}}{\rho}} \quad (2)$$

In Eq. (2), the reduced stiffness is $Q_{22} = \frac{E_{yy}}{1-\upsilon_{xy}^2}$. Results of the fundamental frequencies ($\omega_1^{computed}$) with respect to the material properties are also given in Table 1.

After completing the vibration analysis for the undamaged T-plates, the following experimental analyses are performed for the very early frequency span (0–50 Hz): A. Vibration analysis with four crack-damages with 3 mm in length and 0.5 mm in width, B. Vibration analyses with the four crack-damages and an impact of 0.7 J that causes a 2 mm^2 circular area and a maximum depth of 0.1 mm as shown in Fig. 2.

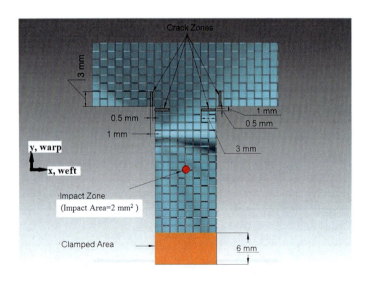

Fig. 2 Geometrical representation of a damaged plain weave T-plate

2.2 Numerical Solution

The general purpose finite element code ANSYS is used for the numerical vibration analyses of the plain weave composite T-plates with cracks and a damage pattern. It is generally known that the most critical part of these analyses is the finite element model which consists of representative volume elements (RVEs). In other words, the RVE plays an important role in the mechanics of composite materials, especially for woven type composites. In this study, a smaller and effective RVE is considered as shown in Fig. 3. Therefore, in order to create a finite element model (FEM) using both a higher order three-dimensional 20-node solid element (SOLID 186) and a three-dimensional 10-node tetrahedral structural solid element (SOLID187), the proposed RVE is taken into account. The representation of the FEM is given in Fig. 4. While a four-ply model includes 1017064 nodes and 140412 elements, a five-ply model includes 1271381 nodes and 175078 elements.

Then a modal analysis is able to be done with respect to the global stiffness matrix $[K]$ and the global mass matrix $[M]$. The first natural frequencies of the composite plates are calculated by the following equation:

$$[K]\{U\} = \omega^2[M\{U\}] \quad (3)$$

In Eq. (3), ω is the natural frequency of the plate considered and $\{U\}$ is the normalized eigenvector. The eigenvalues and eigenvectors are computed by using a preconditioned conjugate gradient Lanczos solver. The mass effect of the accelerometer at the free end is to be taken into consideration. It is generally known that the first two modes are dominated by the accelerometer-mass-effect.

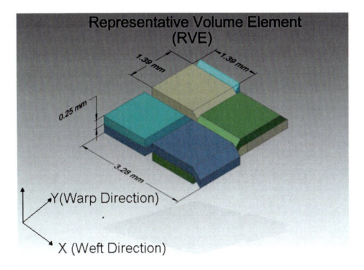

Fig. 3 Representative volume element for the finite element model of the T-plate

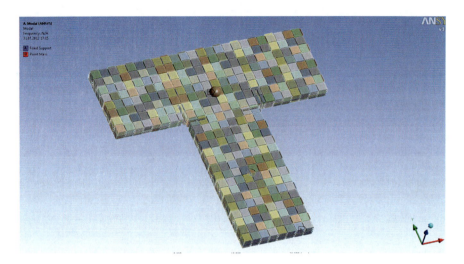

Fig. 4 FEM of the plain weave composite T-plates with cracks and a damage pattern

Therefore, a finite element model check is handled for the mass effect on the undamaged type as shown in Fig. 5. It is obviously seen from the finite element analysis that the attached mass decreases the natural frequencies of the composite plates. In addition to that the first natural frequencies of undamaged composite plates are obtained and given in Table 2. It is clearly seen that the proposed finite element model is able to give acceptable results for the fundamental natural frequencies (i.e., first natural frequencies).

3 Results and Conclusion

From the previous section, the finite element model gives acceptable results for the undamaged T-plates with both 2.548 and 3.06 mm in thickness. First three natural frequencies of the T-plates with A: four crack-damages of 3 mm in length and

Fig. 5 Attached-mass-effect on the first natural frequencies for the composite T-plates

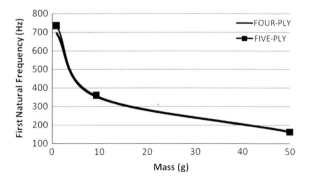

Table 2 Comparison of first natural frequencies of undamaged composite T-plates between the finite element results and the others

Specimen type	$\omega_1^{measured}$ (Hz)	$\omega_1^{computed}$ (Hz)	ω_1^{FEM} (Hz)
Four-ply	347	345.33	353.14
Five-ply	361	365.028	359.02

Table 3 FEM results for first natural frequencies of undamaged (UD), four-crack-damage (A) and four-crack-and-an-impact-damage (B) composite T-plates

Specimen type	UD			A			B		
	Mode I	Mode II	Mode III	Mode I	Mode II	Mode III	Mode I	Mode II	Mode III
Four-ply	353.14	765.52	1931.1	184.68	224.56	927.62	180.48	255.6	886.61
Five-ply	359.02	804.71	1509	158.71	224.65	882.68	165.91	238.58	890.14

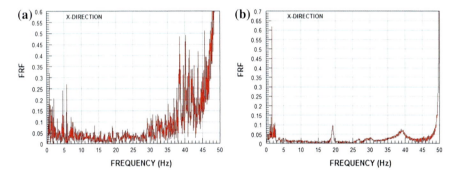

Fig. 6 Frequency response analyses of four-ply T-plate with four cracks before impact (**A**) and after impact (**B**) in x-direction

0.5 mm in width and B: four crack-and-impact-damages including circular impact area of 2 mm² and a depth of 0.1 mm are also obtained from the FEM. Results are given in Table 3.

After completing the vibration analysis for the damaged T-plates, following experimental analyses are performed for the very early frequency span (0–50 Hz). That is, throughout Figs. 6, 7, 8, 9, 10, 11 include not only frequency response analyses (FRF) with four crack-damages (A) but also frequency response analyses with the four crack-damages and an impact of 0.7 Joule (B). For all x-, y- and z-directions, the comparison of A and B situations in Figs. 6, 7, 8, 9, 10, 11 says that the fluctuation of FRF is decreasing while increasing damage effect by the impact of 0.7 J in the frequency span of 0–50 Hz. However, while FRF values are decreasing with increasing the damage effect of 0.7 J in x- and y-directions, FRF values of both A and B are similar corresponding to the thickness direction, i.e., z-direction.

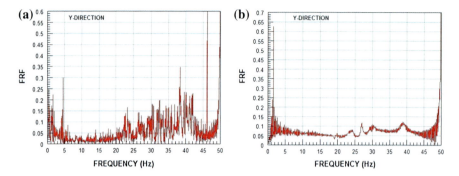

Fig. 7 Frequency response analyses of four-ply T-plate with four cracks before impact (**A**) and after impact (**B**) in y-direction

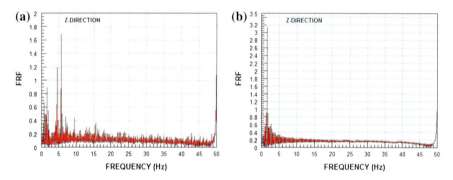

Fig. 8 Frequency response analyses of four-ply T-plate with four cracks before impact (**A**) and after impact (**B**) in z-direction

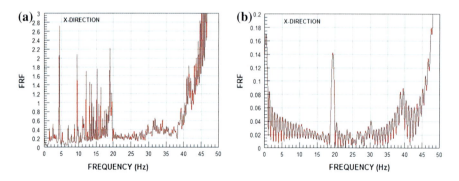

Fig. 9 Frequency response analyses of five-ply T-plate with four cracks before impact (**A**) and after impact (**B**) in x-direction

Vibration Analysis of Carbon Fiber T-Plates

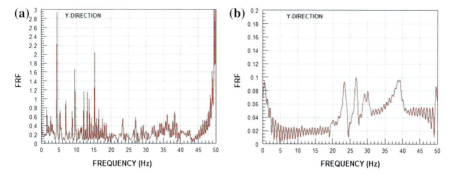

Fig. 10 Frequency response analyses of five-ply T-plate with four cracks before impact (**A**) and after impact (**B**) in y-direction

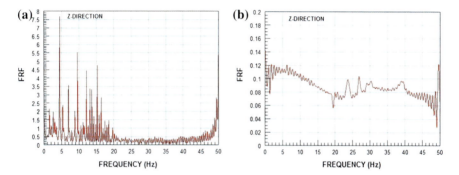

Fig. 11 Frequency response analyses of five-ply T-plate with four cracks before impact (**A**) and after impact (**B**) in z-direction

References

1. Stubbs, N., Osegueda, R.: Global non-destructive damage evaluation in solids. Int. J. Anal. Exp. Modal. Anal. **5**(2), 67–79 (1990)
2. He, J., Fu, Z-F.: Modal analysis. Butterworth-Heinemann Inc., Boston (2001)
3. Deobald, L.R., Gibson, R.F.: Determination of elastic constants of orthotropic plates by a modal analysis Rayleigh-Ritz technique. J. Sound Vib. **124**(2), 269–283 (1988)
4. Salawu, O.S.: Detection of structural damage through changes in frequency: a review. Eng. Struct. **19**(9), 718–723 (1997)
5. Chen, B., Chou, T.-W.: Free vibration analysis of orthogonal-woven fabric composites. Compos. A **30**, 285–297 (1999)
6. Kessler, S.S., Spearing, S.M., Atalla, M.J., Cesnik, C.E.S., Soutis, C.: Damage detection in composite materials using frequency response methods. Compos. B **33**(1), 87–95 (2002)
7. Maia, N.M.M., Silva, J.M.M., Almas, E.A.M., Sampaio, R.P.C.: Damage detection in structures: from mode shape to frequency response function methods. Mech. Syst. Signal. Process. **17**(3), 489–498 (2003)
8. Kim, H.-Y.: Vibration-based damage identification using reconstructed FRFs in composite structures. J. Sound Vib. **259**(5), 1131–1146 (2003)

9. Barbero, E.J., Trovillion, J., Mayugo, J.A., Sikkil, K.K.: Finite element modeling of plain weave fabrics from photomicrograph measurements. Compos. Struct. **73**, 41–52 (2006)
10. ANSYS reference manual, release 12.1 (2012)
11. Jones, R.M.: Mechanics of composite materials, 2nd edn. Taylor & Francis Inc., Philadelphia (1999)

Mechanical Characteristics of AA5083: AA6013 Weldment Joined With AlSi12 and AlSi5 Wires

Mehmet Ayvaz and Hakan Cetinel

Abstract Today, AA5083 and AA6013 aluminum alloys among wrought aluminum alloys are widely used in aerospace, shipbuilding and automotive industries. These aluminum alloys differ from each other from the point of view of weldability, endurance properties and being convenient for heat treatment. AA5083 and AA6013 are welded by the tungsten inert gas welding method with two different electrodes (AlSi12—AlSi5) and six different sample parameters have been obtained as 5083-AlSi12-5083, 5083-AlSi5-5083, 6013-AlSi12-6013, 6013-AlSi5-6013, 5083-AlSi12-6013, 5083-AlSi-6013. The mechanical properties of the samples were investigated by micro-hardness scans, tensile, Charpy impact and three point bending tests.In addition to these, some of the welding samples of 6013-AlSi12-6013, 6013-AlSi5-6013, 5083-AlSi12-6013, 5083-AlSi5-6013 which consist of 6013 alloy were exposed to ageing heat treatment, hardness scans, Charpy impact and three point bending tests. The obtained results have been analyzed and compared to previous results.

Keywords TIG · AA5083 · AA6013 · AlSi12 · AlSi5 · Welding · Three point bending

1 Introduction

Nowadays, aluminum alloys are used in a sort of industries including marine, defense, automotive, transportation, and aerospace [1–8]. Common to all of these industries is the need to weld parts together with fusion based welding processes. Even so, aluminum alloys such as AA7085, AA7040, AA6013, AA5083 and

---

M. Ayvaz · H. Cetinel (✉)
Faculty of Engineering, Department of Mechanical Engineering, Celal Bayar University, Manisa, Turkey
e-mail: hcetinel@cbu.edu.tr

AA2099 have some challenges in welding to other aluminum alloys or to each other [8]. For example, it may be challenging to weld an AA6013 aluminum alloy base metal segment to an AA5083 aluminum alloy to base metal segment, with conventional fusion based welding processes (e.g., gas tungsten arc welding (GTAW-TIG)) because of hot cracking and solidification [1, 8–11]. Therefore, another welding methods involving Friction-Stir Welding (FSW) and laser welding were evaluated and investigated to join dissimilar aluminum alloys.

Friction stir welding is a novel friction-welding process recently developed by TWI (The Welding Institute, Cambridge, U.K.) [3]. Since then, many researchers have explored FSW for welding similar and dissimilar aluminum alloys such as 2017/6013, 2219/5083, 2139/5083, 6061/5083, 6013/6013, 5083/5083 [1, 3, 4, 9, 12–15] and now they know that FSW which does not include melting has the potential to join different and same type of aluminum alloys as hot cracking does not arise [4, 9].

Another solution for joining similar/dissimilar aluminum alloys is the laser welding method which has many advantages such as high welding speed, low distortion, manufacturing flexibility and ease of automation [10]. Because of these advantages, a number of researchers have studied laser welding for welding of similar/dissimilar aluminum alloys [2, 10, 16]. Some of them are AA5083 and AA6013, which have the ballistic resistance and are used in defense and aerospace [17–19]. Broun studied about "Nd:YAG laser butt welding of AA6013 using silicon and magnesium containing filler powders" and figured out that AlSi12 is the most appropriate filler powder to weld AA6013 [10].

Nowadays, AA5083 and AA6013 aluminum alloys are widely used because of their good weldability, corrosion and ballistic resistance but main reasons for the using of these alloys are its lighter weight [1, 3, 7, 17, 18]. For example, because of this reasons AA5083 has been used in the M1113 and the M109 which are military-vehicle systems and age hardenable aluminum alloys, one of the which is AA6013, are important for the aerospace industry [1, 14, 19].

Although, FSW and laser welding are the most efficient methods today, gas tungsten arc welding (GTAW-TIG) is still widely used for joining similar/dissimilar aluminum alloys. However, there is no information in open literature about the TIG welding method to join AA5083-AA6013. Thus, in the present work, mechanical properties of these materials welded with TIG were focused on. Dissimilar aluminum alloys (AA5083-AA6013) and similar aluminum alloys (AA5083-AA5083, AA6013-AA6013) were welded by the TIG welding method and AlSi5 and AlSi12 wires were used to weld these aluminum alloys. Effects of changes of the wires, aluminum alloys and heat treatment on mechanical properties of welded specimens were investigated. As results of tests, it was seen that in the welding of dissimilar aluminum alloys, values of endurance, rigidity and hardness are depend on some factors. These are method and parameters of welding, chemical compositions of base metals (BM) and filler rods, heat treatment and its parameters.

2 Experimental

Herein, aluminum alloys 5083 and 6013 were joined by the TIG welding method with two different filler rods (AlSi12—AlSi5) and six different sample parameters have been obtained as 5083-AlSi12-5083, 5083-AlSi5-5083, 6013-AlSi12-6013, 6013-AlSi5-6013, 5083-AlSi12-6013, 5083-AlSi-6013. Mechanical properties of the samples were investigated by the micro-hardness scans, tensile testing, Charpy and three point bending tests, and the grain structure of the welded samples was examined via optical microscopy. While polishing firstly sandpaper and then diamond was used. After polishing for 20 s, etching was carried out in a 0.5 % aqueous solution of hydrofluoric acid. In addition to these, some of the welding samples 6013-AlSi12-6013, 6013-AlSi5-6013, 5083-AlSi12-6013, 5083-AlSi5-6013 which consist of 6013 alloy have been exposed to age hardening.

The Al-alloy 6013 was developed by Alcoa to take the place of AA 2024-T3. 6013-T6 has 12 % higher tensile strength, 30 % higher compression yield strength, 3 % lower density than 2024-T3 which is the traditional alloy [2, 7]. Therefore, this high-strength aluminum alloy is widely used in aerospace applications. The chemical composition of AA 6013 is given in Table 1.

AA 5083, the other material in this study, has good weldability, corrosion and ballistic resistance properties. Thus, the aluminum alloy 5083 maintains its importance for military industry [14]. The chemical composition of AA 5083 is seen in Table 2.

Because of its convenience, TIG welding is one of the most used methods to join aluminum alloys. Good quality welds, low distortion, free of spatter are some benefits of this method. In TIG welding, an arc is formed between a nonconsumable tungsten electrode and the metal being welded and inert gas such as argon is used for shielding. Two different wires (AlSi12 and AlSi5) were used to join the AA5083 and AA6013 alloys. Welding parameters are shown in Table 3. The chemical compositions of AlSi5 and AlSi12, which was used as filler rods, are given in Table 4.

In the present work, Charpy impact, tensile testing and three point bending tests were made. The dimensions of the samples used in the tensile tests were prepared in accordance with DIN 50120 and tests were performed on a Shimadzu AG-IS (100 kN) testing machine carried out at a cross-head speed of 1 mm/min. The samples used in three point bending tests were prepared in accordance with EN 910/ISO 5173. The test was performed on a Shimadzu Autograph AG (0.5 mm/min displacement, to 180 degrees of bending). The microhardness of the aluminum alloys specimens was measured using a microhardness tester at 100 g for 10 s with a 136° diamond pyramid indenter. For each sample, the microhardness

Table 1 Chemical composition of the 6013 base material

AA	DIN	Fe	Si	Cu	Mn	Mg	Zn	Ti	Cr
6013	AlMg1Si0.8CuMn	0.5	0.6–1.00	0.6–1.10	0.2–0.8	0.8–1.2	0.25	0.10	0.10

Table 2 Chemical composition of the 6083 base material

AA	DIN	Fe	Si	Cu	Mn	Mg	Zn	Ti	Cr
5083	AlMg4.5	0.40	0.40	0.10	0.4–1.00	4.0–4.9	0.25	0.15	0.05–0.25

Table 3 TIG welding parametres

Welding parameters	
Current	Alternative current (70 A)
Gas flow	7 lt/min
Electrode	Tungsten 2.4 mm
Balance	4
Welding speed	3 mm/sec
The shielding gas	Argon

Table 4 Chemical composition of the AlSi12 and AlSi5 electrodes

Filler rods	Si	Mn	Fe	Cu	Mg	Ti	Al
AlSi12	12	0.15	0.60	0.20	–	–	88
AlSi5	5	0.05	0.40	–	0.05	0.15	88

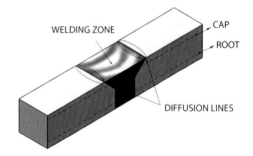

Fig. 1 Figure of used sample in microhardness measurements

measurements were performed for two different sections which were close the bottom (root) and top of the weld zone (cap), as seen in Fig. 1. The aluminum alloys samples were solution treated for 60 min at 530 °C and quenched in water. Artificial aging was carried out at a temperature of 200 °C for 4 h. Then, Charpy impact test, microhardness measurements and three point bending test were performed for the aged samples.

3 Results and Discussion

The microviews of the welded samples are shown in Figs. 2, 3, 4 and 5. In Figs. 2a, b and 3a, b, it can be seen that there are grain orientations in transition zones. The Microstructure of the AlSi12 and AlSi5 welding zones are given in

Figs. 2c and 3c. As can be seen in these figures, the grain size of the AlSi12 welding zone is smaller than the grain size of the AlSi5 welding zone. It is well known that hardness and mechanical properties of the metals increase with decreasing grain size.

The typical dendritic structure (light phases) and eutectic Si particles (dark phases) are shown in Figs. 4 and 5. These figures give information about how the solidification direction is and show how the grain size in cap and root welding zones are. As can be seen, the solidification direction and dendritic orientation are from BM to the weld nugget. In addition to this, Fig. 5 shows the grain size of the cap zone is smaller than grain size of the root zone.

The microhardness variations are shown in Figs. 6, 7, 8, 9, 10 and 11. As can clearly be seen in Figs. 2, 3, 6 and 7, both AlSi5 and AlSi12 weld zones are softer than BM (Base Metal) AA5083 and AA6013. Al-Zn-Mg-Cu, Al-Li, Al-Mg-Si (like 6013) aluminum alloys are known age-hardenable alloys [7, 20], but in contrast the aluminum alloy 5083 (Al-Mg) is in non-age-hardenable aluminum alloys [14, 21]. Reasons of this situation are also seen Figs. 4 and 5. Nonetheless, same figures are interestingly shown, the hardness of the Heat Affected Zone

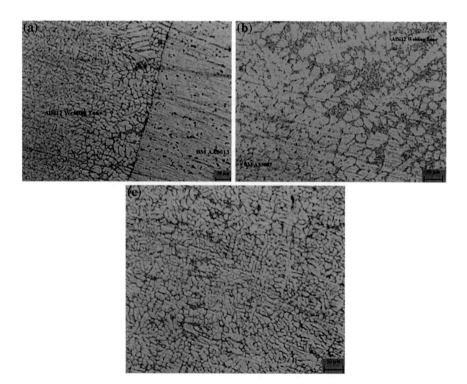

Fig. 2 a Optic micrograph of the transition zone between AlSi12 welding zone and BM AA6012, b optic micrograph of the transition zone between AlSi12 welding zone and BM 5083 and c optic micrograph of the AlSi12 welding zone

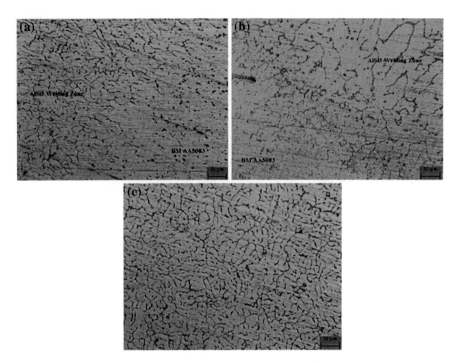

Fig. 3 **a** Optic micrograph of the transition zone between AlSi5 welding zone and BM AA6012, **b** optic micrograph of the transition zone between AlSi5 welding zone and BM 5083 and **c** optic micrograph of structure of the AlSi5 welding zone

Fig. 4 Optic micrograph of the dendritic structure in the AlSi12 welding zone

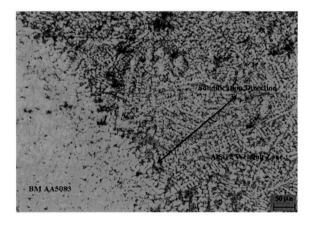

(HAZ), between AA5083 and weld nugget, is slightly higher than its hardness before age-hardening. Likewise, hardness measurements of AlSi5 and AlSi12 weld nuggets are considerably higher than their hardness before age-hardening and the

Fig. 5 Optic micrograph of the solidification direction between cap and root

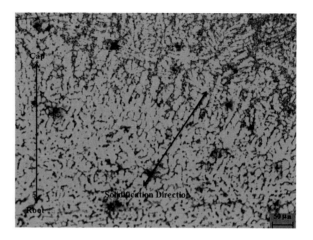

hardness of the BM AA 5083. Besides, hardness rising in AlSi12 is a bit more than in AlSi5.

A slightly increase of hardness value in the HAZ which is between AA5083 and weld center is due to Si diffusion. Another increase of hardness in AlSi5 and AlSi12 weld centers is due to the effect of the Mg interference to nuggets from molten AA5083 (Al-Mg) and AA6013 (Al-Mg-Si). Because of Si diffusion to the HAZ close to AA5083 and Mg mixing to weld zones from base metals when welding, Mg_2Si structures which are the inter-metallic phase were formed during age hardening, as can be seen in the microhardness profiles.

The results of three point bending tests are summarized in Table 5. These results will give information about the rigidity of the samples, when maximum force and maximum displacement of the samples compare. As can be seen, generally, the AlSi12 weld zone is more rigid than AlSi5 weld zone. Additionally,

Fig. 6 Microhardness variation of 5083-AlSi5-5083(root) and 5083-AlSi12-5083(root)

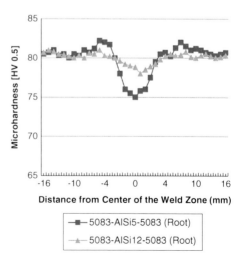

Fig. 7 Microhardness variation of 5083-AlSi5-5083(cap) and 5083-AlSi12-5083(cap)

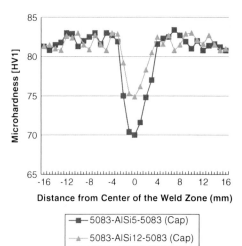

Fig. 8 Microhardness variation of 5083-AlSi5-6013(cap) and 5083-AlSi12-6013(cap)

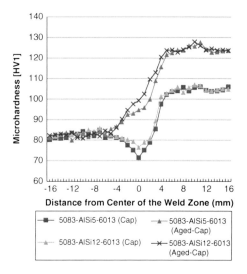

after the age hardening, while maximum forces are rising, values of maximum displacements are on the decline. Therefore we can say that effect of age hardening on the welding zones is seen to increase rigidity and decrease elongation.

In tensile tests, at least three samples were prepared for each different parameter and in the aggregate, 18 samples were tested. The average values of measurements are given in Table 6. As expected, fracture was seen in the weld zones for all samples. It was observed that the AlSi5 weld zone was more ductile than AlSi12's. As can be seen in Table 6, welded dissimilar alloys, 5083-AlSi5-6013 and 5083-AlSi12-6013, had the lowest values. While the minimum tensile strength value is for the 5083-AlSi5-6013's, the 5083-AlSi12-5083 has the maximum tensile strength.

Fig. 9 Microhardness variation of 5083-AlSi5-6013(root) and 5083-AlSi12-6013(root)

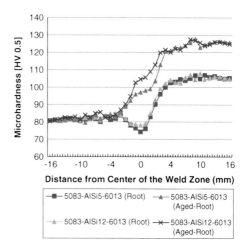

Fig. 10 Microhardness variation of 6013-AlSi5-6013(cap) and 6013-AlSi12-6013(cap)

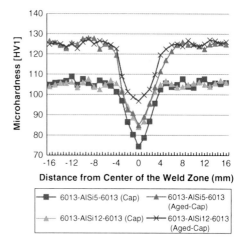

Figures 12, 13 and 14 shows Charpy impact tests results. The toughness of the samples can be evaluated by looking at these figures. Figure 12 indicates that the welding zone and diffusion line of the AA5083-AlSi5-AA5083 have higher impact resistance than AA5083-AlSi12-AA5083 sample's. Besides, as can be seen in Fig. 13, AA6013 samples welded with AlSi5 filler rod have higher impact strength than welded samples with AlSi12. As can be seen in Fig. 13, after age-hardening, fragility increased.

Charpy impact test results of welded dissimilar aluminum alloys, 5083-AlSi5-6013 and 5083-AlSi12-6013, are shown in Fig. 14. From Fig. 14, it can be seen that the 5083 diffusion line is somehow more rigid than the 6013 diffusion line. Moreover, decrease in impact resistance after age-hardening for AlSi12 is higher than for AlSi5.

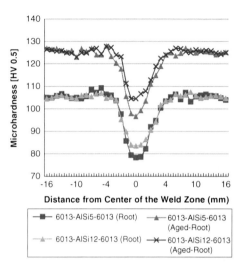

Fig. 11 Microhardness variation of 6013-AlSi5-6013(root) and 6013-AlSi12-6013(root)

Table 5 Result of three point bending tests

Sample	Maximum force (kN)	Maximum displacement (mm)	Damage
5083-AlSi5-5083	1.075	10.18	–
5083-AlSi5-5083	0.82	9.05	–
5083-AlSi12-5083	1.175	7.27	–
5083-AlSi12-5083	1.21	6.048	–
6013-AlSi5-6013	0.6	9.64	
6013-AlSi5-6013	0.85	15.18	–
6013-AlSi12-6013	0.57	9.32	✔
6013-AlSi12-6013	0.61	11.00	–
5083-AlSi5-6013	0.46	7.04	✔
5083-AlSi5-6013	0.76	11.00	–
5083-AlSi12-6013	0.66	11.31	✔
5083-AlSi12-6013	0.92	14.87	–
6013-AlSi5-6013 (Age hardness)	0.86	17.02	✔
6013-AlSi12-6013 (Age hardness)	0.81	9.01	✔
5083-AlSi5-6013 (Age hardness)	0.95	8.35	✔
5083-AlSi12-6013 (Age hardness)	1.17	8.89	–

Table 6 Result of tensile tests

Sample	Yield strength (N/mm^2)	Tensile strength (N/mm^2)	Maximum strain (%)	Fracture from
5083-AlSi5-5083	132.17	161.369	10.33	Welding
5083-AlSi12-5083	138.738	196.035	10.46	Welding
6013-AlSi5-6013	132.65	145.034	11.36	Welding
6013-AlSi12-6013	125.78	135.156	7.79	Welding
5083-AlSi5-6013	120.52	123.272	8.76	Welding
5083-AlSi12-6013	124.65	131.890	8.90	Welding

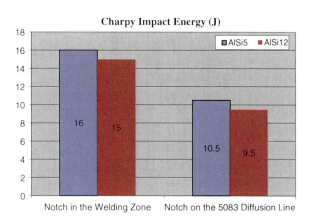

Fig. 12 Charpy impact tests results of 5083-AlSi5-5083 and 5083-AlSi12-5083

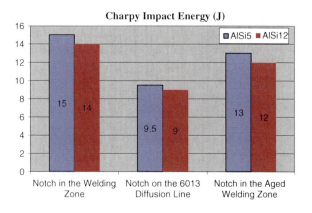

Fig. 13 Charpy impact tests results of 6013-AlSi5-6013 and 6013-AlSi12-6013

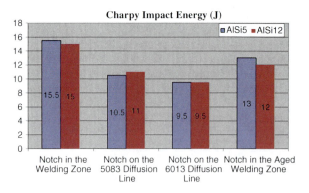

Fig. 14 Charpy impact tests results of 5083-AlSi5-6013 and 5083-AlSi12-6013

4 Conclusion

The most important results are given below:

- The hardness value of the AlSi12 welding zone is rather higher than the AlSi5 welding zone's. And so, in terms of hardness, AlSi12 is somehow nearer than AlSi5 to the base metals, AA5083 and AA6013.
- Normally, AA5083 is a non-age-hardening. However, herein, because of the Si diffusion from the welding zone, after age-hardening, Mg_2Si inter metallic phases were formed in the HAZ which is the near BM AA5083. Therefore, the hardness voule increased in the AA5083 HAZ.
- Due to mixing of base materials and filler rods in the nugget zone, age hardening was possible. This situation has been also proven in Charpy impact and microhardness tests. And, after age-hardening, both hardness values of AlSi5 welding zone and AlSi12 welding zone's passed to AA5083 hardness.
- With age-hardening, hardness values of the BM AA6013 have rised.
- As can be seen in the three point bending tests, after age-hardening, the rigidity of welded samples has rised.
- While the lowest impact resistance is seen in the diffusion line which is between BM AA6013 and welding zone AlSi12, AlSi5 welding zone has the highest impact resistance.

References

1. Kafali, H., Ay, N.: Mechanical properties of 6013-T6 aluminium alloy friction stir welded plate. 13th International Conference, Aerospace Sciences & Aviation Technology (2009)
2. Xu, W.L., Yue, M.T., Man, H.C.: Nd:YAG laser surface melting of aluminium alloy 6013 for improving pitting corrosion fatigue resistance. J. Mater. Sci. **43**, 942–951 (2008)
3. Heinz, B., Skrotzki, B.: Characterization of a friction-stir-welded aluminum alloy 6013. Metall. Mater. Trans. B **33B**, 489–498 (2002)

4. Zhou, C., Yang, X., Luan, G.: Effect of kissing bond on fatigue behavior of friction stir welds on Al 5083 alloy. Mater. **41**, 2771–2777 (2006)
5. Lenczowski, B.: New lightweight alloys for welded aircraft structure. ICAS2002 Congress (2002)
6. Barbosa, C., Rebello, J.M.A., Acselrad, O., Dille, J., Delplancke, J.-L.: Identification of precipitates in 6013 aluminum alloy (Al-Mg-Si-Cu). Z. Metallkd. **93**, 208–211 (2002)
7. Varlı, A.E., Gurbuz, R.: Fatigue crack growth behaviour of 6013 aluminium alloy at different ageing conditions in two orientations. Turk. J. Eng. Env. Sci. **30**, 381–386 (2006)
8. Lin, C.J., Stol, I., Williams, K.L.: (2010) Fusion weldable filler alloys. United States Patent Application Publication Lin et al. Pub. No.: US 2010/0129683 A1
9. Dilip, J.J.S., Koilraj, M., Sundareswaran, V., Janaki Ram, G.D., Koteswara Rao, S.R.: Microstructural characterization of dissimilar friction stir welds between AA2219 and AA5083. Trans. Indian Inst. Met. **64**, 757–764 (2010)
10. Braun, R.: Nd:YAG laser butt welding of AA6013 using silicon and magnesium containing filler powders. Mater. Sci. Eng. A-Struct. **426**, 250–262 (2006)
11. Zhan, B.S., Zhang, Y.M.: Welding aluminum alloy 6061 with opposing dual torch GTAW process. Weld J. 202–206 (1999)
12. Shigematsu, I., Kwon, Y.-J., Suzuki, K., Imai, T., Saito, N.: Joining of 5083 and 6061 aluminum alloys by friction stir welding. J. Mater. Sci. Lett. **22**, 353–356 (2003)
13. Mrozcka, K., Dutkiewicz, J., Lityñska-Dobrzyñska, L., Pietras, A.: Microstructure and properties of FSW joints of 2017A/6013 aluminium alloys sheets. Arch. Mater. Sci. Eng. **33**, 93–96 (2008)
14. Grujicic, M., Arakere, G., Yen, C.-F., Cheeseman, B.A.: Computational investigation of hardness evolution during friction-stir welding of AA5083 and AA2139 aluminum alloys. J. Mater. Eng. Perform. **20**, 1097–1108 (2011)
15. Bradley, G.R., James, M.N.: Geometry and microstructure of metal inert gas and friction stir welded aluminium alloy 5383-H321 (2000)
16. Okon, P., Dearden, G., Watkins, K., Sharp, M., French, P.: Laser welding of aluminium alloy 5083. 21st international congress on applications of lasers and electro-optics, Scottsdale (2002)
17. Newbery, A.P., Nutt, S.R., Lavernia, E.J.: (2006) Multi-scale Al 5083 for military vehicles with improved performance. Jom-J. Min. Met. Mat. S. 56–61
18. Stol, I.: Consepts for weldable ballistic products for use in weld field repair and fabrication of ballistic resistant structure. United States Patent Application Publication. Pub. No.: US 2009/0151550 A1 (2009)
19. Starke, E.A., Jr, Staley, Staley, J.T.: Application of modern aluminum alloys to aircraft. Prog. Aerosp. Sci. **32**, 131–172 (1996)
20. Ringer, S.P., Hono, K.: Microstructural evolution and age hardening in aluminium alloys: atom probe field-ion microscopy and transmission electron microscopy studies. Mater. Charact. **44**, 101–131 (2000)
21. Zhou, C., Yang, X., Luan, G.: Fatigue properties of friction stir welds in Al 5083 alloy. Scripta Mater. **53**, 1187–1191 (2005)

Numeric Simulation of the Penetration of 7.62 mm Armour Piercing Projectile into Ceramic/Composite Armour

Ömer Eksik, Levent Turhan, Enver Yalçın and Volkan Günay

Abstract A numerical study has been carried out to design armour system which consists of 99.5 % alumina ceramic tiles backed by plain weave S2-glass/epoxy laminated composite plates when subjected to high velocity impact of a 7.62 mm armour piercing (AP) projectile. The explicit three-dimensional finite element code LS-DYNA was used in the analysis of the problem. A plastic-kinematic hardening model; MAT-03, the Johnson-Holmquist-Ceramic model; MAT-110 and newly developed composite damage model namely MAT-162 implemented in LS-DYNA were used in order to simulate behaviours of projectile, ceramic and composite layers respectively in the numerical model. Three configurations with different composite backing plate thicknesses were considered in the numerical analyses in order to defeat a 7.62 mm AP projectile. Numerical results showed that among the investigated armour panels the best multi-hit ballistic performance was attained with configuration 3 which maintained ballistic protection with a minimal damage to the composite backing and much smaller back face signature.

Ö. Eksik (✉) · L. Turhan · E. Yalçın · V. Günay
TÜBİTAK Marmara Research Centre, PO Box 41470 Gebze, Kocaeli, Turkey
e-mail: omer.eksik@tubitak.gov.tr

L. Turhan
e-mail: levent.turhan@tubitak.gov.tr

E. Yalçın
e-mail: enver.yalcin@tubitak.gov.tr

V. Günay
e-mail: volkan.gunay@tubitak.gov.tr

1 Introduction

Armour systems such as hard faced ceramic with composite backing are necessary and widely used to defeat AP projectiles. The function of the ceramic facing is to destroy the tip of the incoming projectile, to distribute the load over a large area of composite backing plate and decelerate the projectile. Alumina (Al_2O_3), boron carbide (B_4C), boron silicon carbide (BSC), and silicon carbide (SiC) are some of the ceramics that are commonly used. The composite backing plate absorbs the kinetic energy of the decelerated projectiles and also catches the ceramic and projectile fragments preventing them doing further harm. The range of composite material includes aramid woven fabrics such as Kevlar and Twaron and fibre glass materials such as S2-glass and E-glass. Compared with the traditional metal armours, ceramic/composite armours have high ballistic performance and lightweight [1].

Many researchers have studied ballistic performance of ceramic/composite armours. Krishnan et al. [1] studied the ballistic performance of ceramic/composite armour systems numerically. They developed a material model to observe the damage propagation at the composite backing plate made of ultra high molecular weight polyethylene. They found that the finite element prediction of damage were in good agreement with the experimental data. Turhan et al. [2] performed depth of penetration (DOP) tests and ballistic impact of 7.62 and 12.7 mm AP onto ceramic/aluminium armour systems. Based on the DOP tests results, projectile and aluminium erosion strains were found and used at numerical simulation of ballistic impact tests. They found that results from both experiments and numerical simulations were in good agreement. Fawaz et al. [3] performed numerical simulations of normal and oblique ballistic impact on ceramic/composite armours. The finite element results were compared with experimental data. The findings showed the distributions of global, kinetic, internal and total energy versus time are similar for normal and oblique impact, but the interlaminar stresses at the ceramic/composite interface and the forces at the projectile erosion in oblique impact was slightly greater than for normal impact. Madhu et al. [4] performed ballistic tests on high purity alumina ceramic tiles backed by thick aluminium plates. It was found that high purity alumina shows higher ballistic performance. Mahfuz et al. [5] performed finite element analysis of an integral armour under high velocity impact, in order to assess the in-plane transverse and interlaminar stresses for failure analysis. They found that the correlation with the experimentally observed damage zone was in good agreement with the finite element analysis. Lee and Yoo [6] investigated ballistic efficiency of ceramic/metal armour systems against 7.62 mm AP projectile numerically and experimentally. They used a smoothed particle hydrodynamics scheme in order to capture projectile eroding, formation of ceramic conoid and failure of the backing plate. The results from numerical modelling were in a good agreement with experimental ones. Navaro et al. [7] designed ceramic faced armours backed by composite plates to defeat 7.62 mm AP projectile by an experimentally based on the energy absorption model. Ballistic limit predicted from their model was found to be in good agreement with the experimental data. It was mentioned that [8] woven composites by nature are ideal candidates for use in high performance ballistic and military applications. Woven fibre epoxy composites combine the beneficial properties of both polymer resin (ability to absorb and mitigate kinetic energy) and high performance fibres (high to ultra high elastic modulus). Most importantly they are capable of providing equivalent ballistic protection at a significantly reduced areal weight when compared with traditional metal based armour systems.

In the current study three sets of finite element models have been developed to design of ceramic/composite armours to defeat 7.62 mm AP projectiles. The materials used were 100Cr6, Alumina and S2-glass/SC-15 epoxy composite. 100Cr6 as projectile, Alumina as frontal plate and S2-glass/SC-15 epoxy composite for the backing plate are chosen.

2 Numerical Modelling

In this work, design of ceramic/composite armours were investigated by using explicit dynamic finite element analyses (FEA) code LS-DYNA. The hardened steel core of the projectile geometry was obtained with the ATOS digitizing device as shown in Fig. 1.

The boundary of all targets specifically was chosen to be circular instead of other shapes such as rectangular. This is because of the symmetry of the stress wave propagation and reflection in the circumferential direction of the target that can be preserved in a circular shape but not in rectangular shape during the normal impact of the 7.62 mm AP projectile. Three dimensional finite element meshes for the target and projectile were created by using the ANSYS-LS-DYNA pre-processor which was shown in Fig. 2. The circular plate was divided into two regions in the mesh in radial direction, i.e. in the inner and outer direction. The mesh was

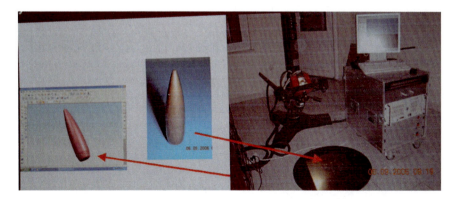

Fig. 1 Geometry of the 7.62 mm AP projectile was obtained by using ATOS digitizing device

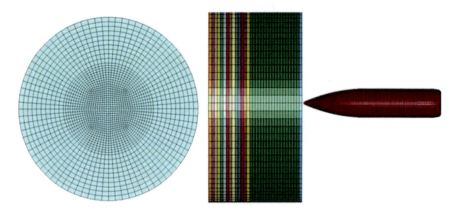

Fig. 2 Finite element model of projectile and armour panel

coarsening from the inner to outer region. The target and projectile were meshed with explicit 8-noded solid elements (SOLID164) with one point integration. The nodes making up the projectile's mesh were assigned an initial velocity. The translational nodal degrees of freedom along the boundary of the target are constrained to prevent any translational movement. The interface between the projectile and armour material is defined with CONTACT-ERODING-SURFACE-TO-SURFACE. Projectile erosion is one of the major features of the armour penetration process. An erosion algorithm is available in the LS-DYNA code which allows penetration and perforation by eroding elements from projectile surface as well as the target structure. In the numerical analyses, ceramic armour with three different composite backing plate thicknesses were considered for the normal impact of the 7.62 mm AP projectile. FEA were performed on a Sun Ultra 40 Workstation with 2 × AMD Opteron 2218 processors and analysis took around 1 h.

3 Material Model and Properties

The utilization of modelling and simulation tools for the design of armour systems is critically based on material models which should accurately reflect the physical behaviour of the armour systems. Two components, namely the ceramic plate and a composite plate are used in the armour system. The projectile is modelled with material type 3 MAT-PLASTIC-KINEMATIC. This is a bilinear elastic–plastic model that contains a formulation combining isotropic and kinematic hardening based on five material properties. The ceramic is modelled with material type 110 MAT-JOHNSON-HOLMQUIST-CERAMIC. Johnson Holmquist-2 material model allows progressive damage modelling of ceramic material. Composite backing plate is modelled with type 162 COMPOSITE-DMG-MSC. Type 162 is a progressive composite damage model and implemented in LS-DYNA. It requires a

Table 1 Plastic kinematic hardening constants for projectile hard steel cores [2]

Modulus of elasticity E (MPa)		205.000
Density ρ_0 (t/mm^3)		7,85E−9
Poisson's ratio ν		0,3
Yield stress σ_y (MPa)		1.500
Tangent modulus E_T (MPa)		670
Strain parameter	C	40
	P	5
Failure strain ε_f (%)		3

total of 34 material properties and computational modelling parameters to describe the response of orthotropic unidirectional and woven composites [9]. Detailed descriptions of these material models can be found in LS-DYNA theoretical manual [10]. The corresponding material properties required for the models are taken from the literature [2, 9, 11] and shown in Tables 1, 2 and 3.

4 Numeric Simulation Results and Discussion

Ballistic simulations of normal impact of a 7.62 mm AP projectile onto ceramic/composite armour for three different composite backing plate thicknesses were simulated. Table 4 and Fig. 3 show the constituents and numerical model of all

Table 2 MAT162 material properties & parameters for plain-weave S-2 Glass/SC-15 composites [9]

MID	ρ (t/mm^3)	E_1 (MPa)	E_2 (MPa)	E_3 (MPa)	ν_{21}	ν_{31}	ν_{32}
162	1,85E−9	27.500	27.500	11.800	0,11	0.18	0.18

G_{12} (MPa)	G_{23} (MPa)	G_{31} (MPa)					
2.900	2.140	2.140					

S_1^T (MPa)	S_1^C (MPa)	S_2^T (MPa)	S_2^C (MPa)	S_3^T (MPa)	S_{FC} (MPa)	S_{FS} (MPa)	S_{12} (MPa)
604	291	604	291	58	850	300	75

S_{23} (MPa)	S_{F1} (MPa)	S_{FFC} (MPa)	ϕ (deg)	e_{limit}	S_{delam}		
58	58	0.3	10	0.2	1.2		

w_{max}	e_{crush}	e_{Expn}	C_{rate1}	m_1			
0,999	0,001	4,5	0,03	2			

m_2	m_3	m_4	C_{rate2}	C_{rate3}	C_{rate4}		
2	0,5	0,2	0	0,03	0.03		

Table 3 Johnson Holmquist-2 constants for AD-99.5 Alumina [11] plastic kinematic hardening constants for projectile hard steel cores [2]

Density ρ_0 (t/mm^3)	3,89E−09
Shear modulus G (MPa)	152000
Hugoniot elastic limit HEL	6570
Intact strength constant A	0,88
Intact strength constant N	0,64
Strain rate constant C	0,007
Fracture strength constant B	0,28
Fracture strength constant M	0,6
Maximum fracture strength SFMAX	1
Tensile strength T (MPa)	262
Pressure constant K1 (MPa)	231000
Pressure constant K2 (MPa)	−160000
Pressure constant K3 (MPa)	2774000
Bulking constant BULK	1
Damage constant (D1)	0.01
Damage constant (D2)	0.7

three configurations. The facesheet in front of the ceramic plate provides through thickness the ceramic confinement during the ballistic impact. The composite backing plate of the armour systems is divided into 14, 19 and 24 layers and one solid element was used to model each layer.

Figure 4 depicts the projectile penetration processes in a ceramic/composite armour system for all three configurations. For a normal impact at 50 μs Figs. 4a–c show that the projectile penetrates to the composite backing plate for all configurations in different penetration thickness (15.7 mm for configuration 1, 13.4 mm for configuration 2 and 12 mm for configuration 3). Figure 4d–f show the final snapshot of the armour panels while the projectile is damaged as it goes through the facesheet, ceramic plate and composite backing plate. Perforation of configuration 1 by the AP projectile occurred while for the configurations 2 and 3 the projectile stops at different penetration thicknesses. The projectile gets eroded and blunted during the penetration process and its length is reduced to 10.5 and 9.82 mm. The projectile erosion in configuration 3 is slightly greater than in

Table 4 Material constituents of the armour panels

Armor panel constituents			
	Facesheet	Ceramic plate (mm)	Composite backing plate
Configuration 1	1 layer S2-Glass/SC-15 Epoxy	10 mm Alumina Ceramic	14 layer S2-Glass/SC-15 Epoxy
Configuration 2	1 layer S2-Glass/SC-15 Epoxy	10 mm Alumina Ceramic	19 layer S2-Glass/SC-15 Epoxy
Configuration 3	1 layer S2-Glass/SC-15 Epoxy	10 mm Alumina Ceramic	24 layer S2-Glass/SC-15 Epoxy

Numeric Simulation of the Penetration

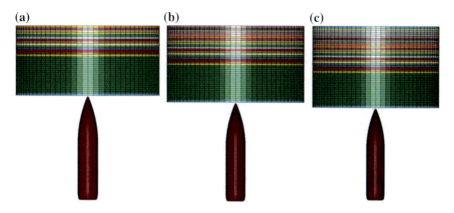

Fig. 3 Numerical model **a** Configuration 1, **b** Configuration 2, **c** Configuration 3

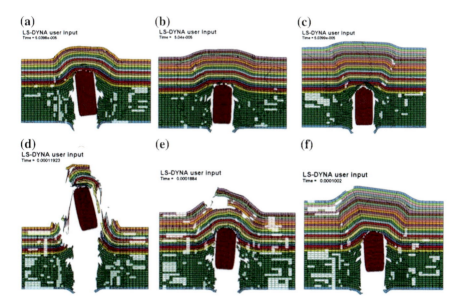

Fig. 4 State of projectile at different times, **a** Configuration 1 at time 50 μs, **b** Configuration 2 at time 50 μs, **c** Configuration 3 at time 50 μs, **d** Configuration 1 at final snapshot, **e** Configuration 2 at final snapshot, **f** Configuration 3 at final snapshot

configuration 2 because of the penetration path in this case is 3 mm longer than that in configuration 2.

Figure 5 shows the back face signature of the configurations 2 and 3 at the time of projectile velocity reduced to zero (maximum dynamic displacement). Less damage was observed in configuration 3 compared to configuration 2.

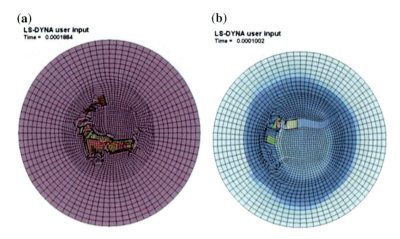

Fig. 5 Back face signature of the armour at final snapshot, **a** Configuration 2 and, **b** Configuration 3

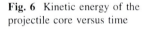

Fig. 6 Kinetic energy of the projectile core versus time

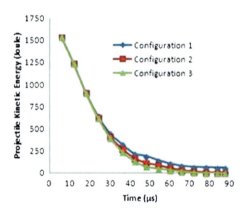

The evolutions of the kinetic energy of the projectile for all three configurations are shown in Fig. 6. In configuration 1 87 % of the kinetic energy of the projectile core is dissipated as it goes through the ceramic layer while 93 and 97 % kinetic energy dissipation of the projectile take place through configurations 2 and 3 respectively.

Figure 7 shows the distribution of the global kinetic, internal and total energy with respect to time for all three configurations. Since the full penetration occurred in configuration 1, 95 % of the kinetic energy of the projectile was dissipated while the kinetic energy of the projectile reduced to zero for the other two configurations. Although the different composite backing plate thickness between configurations 2 and 3, their distributions of global energy are quite similar. For

Fig. 7 Global energy versus time, **a** Configuration 1, **b** Configuration 2, **c** Configuration 3

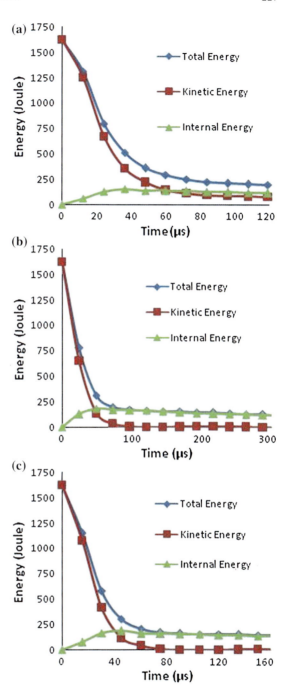

configuration 2 the reduction of kinetic and total energy is about 20 µs slower than that for configuration 3.

All the results showed that an increased backing plate thickness from 14 layers to 19 and 24 layers helps to absorb the residual kinetic energy of the projectile and eventually stops it.

5 Conclusion

A finite element model of the normal impact of a 7.62 mm AP projectile onto ceramic/composite armour was developed to design armour plates. Configuration 3 is chosen to be the best design because the reduction of the kinetic energy of the projectile is quicker which leads to minimal damage to composite backing and much smaller back face signature. Experimental verification of such a model could be used for a large number of threats for designing cost effective armour systems without performing expensive experimental procedures.

References

1. Krishnan, K., Sockalingam, S., Bansal, S., Rajan, S.D.: Numerical simulation of ceramic composite armor subjected to ballistic impact. Compos. B **41**, 583–593 (2010)
2. Turhan, L., Eksik, Ö., Yalçın, E., Demirural, A., Baykara, T., Günay, V.: 10th International Conference on Structures Under Shock and Impact, Algarve, Portugal, WIT Transactions on the Built Environment vol. 98, pp. 379–388 (2008)
3. Fawaz, Z., Zheng, W., Behdinan, K.: Numerical simulation of normal and oblique ballistic impact on ceramic composite armours. Compos. Struct. **63**, 387–395 (2004)
4. Madhu, V., Ramanjanneyulu, K., Bhat, T.B., Gupta, N.K.: An experimental study of penetration resistance of ceramic armour subjected to projectile impact. Int. J. Impact. Eng. **32**, 337–350 (2005)
5. Mahfuz, H., Zhu, Y., Haque, A., Abutalib, A., Vaidya, U., Jeelani, S., Gama, B.A., Gillespie, J., Fink, B.: Investigation of high-velocity impact on integral armor using finite element method. Int. J. Impact. Eng. **24**, 203–217 (2000)
6. Lee, M., Yoo, Y.H., Analysis of ceramic/metal armour systems. Int. J. Impact. Eng. **25**, 819–829
7. Navaro, C., Martinez, M.A., Cortes, R., Galvez, V.S.: Some observations on the normal impact on ceramic faced armours backed by composite plates. Int. J. Impact. Eng. **13**, 145–156 (1993)
8. Grogan, G., Tekalur, S.A., Shukla, A., Bogdanovich, A., Coffelt, R.A.: Ballistic resistance of 2D and 3D woven sandwich composites. J. Sandwich. Struct. Mater. **9**, 283–302 (2007)
9. Xiao, J.R., Gama, B.A., Gillespie, J.W.: Progressive damage and delamination in plain weave S-2 glass/SC-15 composites under quasi-static punch-shear loading. Compos. Struct. **2**, 182–196 (2007)
10. LS-DYNA Theoretical manual. Livermore Software Technology Corporation, Livermore (1998)
11. Anderson, C.E., Johnson, G.R., Holmquist, T.J.: Ballistic performance and computations of confined A1203 ceramic. In: 15th International Symposium on Ballistics, vol. 2, Jerusalem, Israel, May 21–24, pp. 65–72 (1995)

In-situ TEM Observation of Deformations in a Single Crystal Sapphire During Nanoindentation

Fathi ElFallagh, Aiden Lockwood and Beverley Inkson

Abstract Real time nanodeformation processes of a single crystal alumina ($1\bar{1}02$) were studied by in situ Transmission Electron Microscopy (TEM). Tungsten tips of 50 and 450 nm diameters were used to indent the sample. Crack development was observed and analyzed in real time. TEM analysis of the damage area beneath the nanoindentation sites in alumina reveals complex zones of cracks and varying local strains. Cracks with variable sizes were observed some have preferred orientations along different crystal planes, for example {$\bar{1}012$}. Bend contours, due to stress, were observed to develop prior to crack development during the loading cycle of the indenter. Cracks were observed to follow bend contours with initial speed of 17 nm/s. Median cracks were observed to form during loading process of indentation and to propagate parallel to indentation direction.

1 Introduction

Most of the previous work on the deformation of alumina with indentation was post-mortem. In-situ nanoindentation in TEM is a recent development in the field [1, 2]. Special TEM nanoindentation holders have been developed, including one developed in our group [3, 4]. Micro-and nano-indentation of brittle material generates a significant number of cracks which nucleate and propagate to relieve

F. ElFallagh (✉)
Azzaytuna University, Soog Alahad, Tarhuna, Libya
e-mail: Fathi_Fallagh@yahoo.com

A. Lockwood · B. Inkson
Department of Engineering Materials, The University of Sheffield,
Mappin Street, Sheffield S1 3JD, UK
e-mail: a.lockwood@sheffield.ac.uk

B. Inkson
e-mail: Beverley.inkson@sheffield.ac.uk

the high local stresses generated during the indentation process [5–9]. In the last 10 years it has been realised that the microstructural deformation and induced cracks around indents can be conveniently studied by cross-sectioning the indent by localised sputtering using a highly focused ion beam (FIB) [10–21]. Alumina is an extremely important wear resistant material, and its surface mechanical properties have been studied for many decades [5, 7–9, 22–29]. Most microstructural studies of alumina wear have involved sectioning wear surfaces post-mortem into 100–1000 nm thick slices and examining the retained microstructure and dislocation distribution by TEM (e.g. [7, 9, 26, 27]). Studying the deformations of thin alumina samples in real time offers valuable information about the dynamic behaviour of alumina at the nanoscale when subjected to stress. The elastic and the plastic deformation of alumina can be studied by in situ TEM. Dislocation initiation and movements along with crack development and dynamics can be all studied in real time. In this study the deformation of a single crystal alumina was studied in real time by nanoindentation with tungsten tips using a special in situ TEM holder.

2 Experimental Procedures

Sapphire samples were ground from the original thickness of 0.55 mm to 50 μm in steps using a Minimet 1000 grinding and polishing machine with 30, 15 and 6 μm diamond grit and polished using a 1 μm diamond film. Samples were mounted on 3 mm TEM copper disks using epoxy glue with one edge of the sample at the centre as shown in Fig. 1. Using a Precision Ion Polishing System (PIPS 691) at 5 kV and polishing angle of 5 degrees TEM samples were sputtered to electron transparency <1 μm.

The sample was then mounted in the in situ TEM holder fitted with a sharp tungsten tip (see Fig. 2; [3]).

The alumina sample was fitted on the in situ TEM holder designed by our nanorobotics group [3, 4, 30]. The holder was built to fit the JEOL 2010 and JEOL

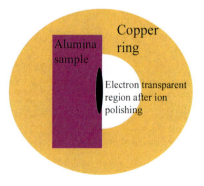

Fig. 1 Schematic of alumina sample fitted to 3 mm copper disk, showing polished region by PIPS

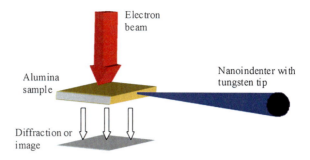

Fig. 2 Schematic of in situ TEM nanoindentation of alumina

3000F (Japan) series microscopes. The holder has the external geometry of the standard TEM holders but from the inside it was fitted with mechanisms that generate coarse and fine displacements in the range mm → 0.1 nm. A 3-axis coarse positioner based on inertial sliders is used to get the sharp tip close to the sample within the range of the fine positioner. The fine positioner of the nanoindenter consists of a four quadrant piezoelectric tube (PZT; diameter 6 mm). The localized deformation in the sample is induced by pushing the tip (speed of 1.6 nm/s) into the sample with the help of this fine positioner [3]. Video tapes of the deformation process and diffraction patterns were recorded in situ during the deformation of the alumina sample by the tungsten tip.

3 Results and Discussion

3.1 Real-Time Fracture of R-Cut Alumina During Deformation by Nanoindentation

Diffraction patterns were recorded in situ during the deformation of the alumina sample by the tungsten tip ~50 nm in diameter (see Fig. 3). At the beginning the sample orientation was in the zone axis [2 $\bar{2}$0 1] (see Fig. 4) then the sample was oriented so that the electron beam was parallel to the ($\bar{1}$104) plane and as the indenter started to bend the sample it can be seen that different planes are becoming parallel to the electron beam (Fig. 3a–c). Some diffraction spots from the tungsten tip became visible in Fig. 3c–i along with the diffraction spots from the alumina as the tip moved into the selected area. As the indenter tip was retracted the diffraction spots from the tungsten disappeared in Fig. 3j–l and the diffraction from the alumina became prominent. The sample exhibited elastic behaviour and returned to the original orientation.

Figure 5 shows six video frames extracted from a 37 s deformation sequence for the R-cut alumina sample. The sharp tip, with speed of 1.6 nm/s contacted the sample at the edge and overlapped the sample by ~400 nm (Fig. 6.). Initially the alumina sample elastically deformed and clear bend contours were formed

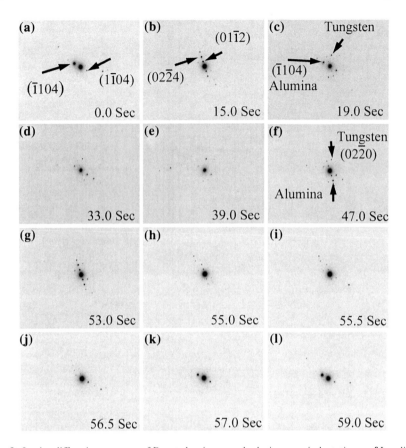

Fig. 3 In-situ diffraction patterns of R-cut alumina sample during nanoindentation. a–f Loading. g–l Unloading

(Fig. 5a) radiating from contact point. Some of these contours have dark contrast while others appear bright due to diffraction induced electron scattering. Each bend contour corresponds to an area of constant diffraction condition. As the single crystal sample initially had a constant orientation, the bend contours correspond to increase of stress within the sample [2].

The bend stress contours formed curves around the location of the tip, as can be seen in Fig. 5, and moved away from the contact point with increasing strain. Their speed was relatively fast at the beginning of the experiment. As time increases their speed decreases, and their shape changed from wide open curves to a hyperbolic shape and start to intersect, as can be seen in Fig. 5c, d. A crack initiated at t ~ 31.5 s. and was observed to propagate into the specimen starting from the intersection point of the contours, see Fig. 5d. The crack followed one of the bend contours to the edges of the sample (Fig. 5e, f). Figure 7 depicts the process of the crack development during the indentation, the crack initiates at the contact point under the indenter and follows path 1 with relatively low speed

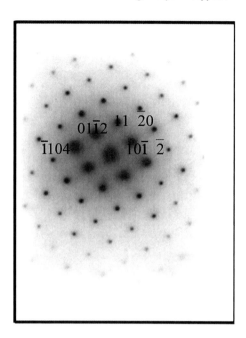

Fig. 4 Electron diffraction patterns from the edge of the alumina sample, showing [2 0$\bar{1}$I2 1] zone axis

(17 nm/s). Then it follows paths 2 and 3 in the same time with higher speed resulting into two material fragments, the fragment between crack 1 and 2 falls off the specimen and the second fragment between 1 and 3 stay attached to the bulk specimen as can be seen in Fig. 5e, f, the depth of the indenter at this stage of the process was ∼ 500 nm. The initial crack length (see Fig. 7 path 1) starting from the estimated contact point to the intersection point between paths 1, 2 and 3 was measured as function of time and plotted in Fig. 8. The slope of the line in Fig. 8 was found to be 17 nm/s which is the initial speed of the crack (path 1). The circumferential crack that developed was measured to be ∼ 660 nm in length (path 2 and 3) and in part could be following the ($\bar{1}$012) plane. The fact that fracture occurs preferentially along the ($\bar{1}$012) plane agrees with the observation of Stofel and Conrad [31, 32] noted fracture along the {$\bar{1}$012} planes in bend tests.

3.2 Fracture of R-Cut Alumina During Nanoindentation

First trials to indent the alumina sample with a sharp 50 nm tungsten tip on the thin edge (electron transparent region) <40 nm in thickness (see Fig. 6) generated significant bending due to the flexibility of the free edge of the sample. In the

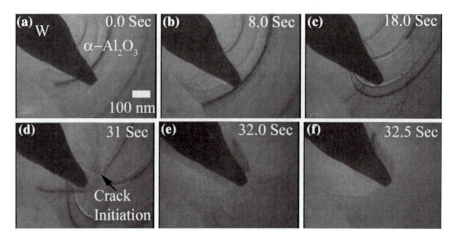

Fig. 5 a–f In-situ fracture sequence during deformation of R-cut alumina sample with a 50 nm radius tungsten nanoindenter tip (time of video frame indicated)

Fig. 6 Schematic showing contact point of indenter with the sample

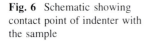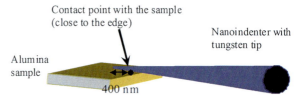

following experiment a larger tungsten tip with ~450 nm in diameter was used to indent a thicker region of the sample.

The indentation process was recorded in situ on video for analysis. Figure 9 shows six video frames extracted from a 2:55 min deformation sequence. During first contact of the indenter with the sample, bend contours in the shape of open curves were formed around the contact point (see Fig. 9a) at 14.0 s (Fig. 9b). In the loading process, a median crack was observed to develop and propagate parallel to the direction of loading and deviates side way with the increasing time of loading (see Fig. 9c).

Figure 10 shows a graph of the displacement of the indenter tip with time, the applied displacement is linear, however the measured displacement fluctuates due to the material in contact. Displacement bursts are seen as the material cracks over several stages (see Fig. 10). The relative speed of the tip, during contact with the sample, was calculated from the graph to be 0.27 nm/s. Figure 11 shows the centre of the indentation site with damage, median and lateral cracks were observed. The median crack (with length >500 nm and width <10 nm) was parallel to the direction of indentation. Stress contours were observed during and after indentation (Figs. 9, 11). Some of these contours terminate at the median cracks (see Figs. 9b, 11b) their shape is semi-elliptical and occur in pairs most of the time

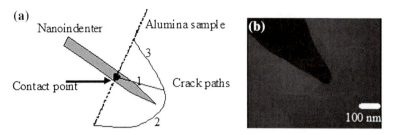

Fig. 7 a Schematic of the process of crack development during in situ nanoindentation. **b** In-situ TEM image of crack development

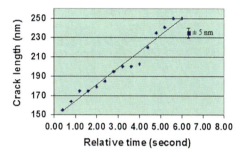

Fig. 8 Graph of the crack length versus relative time

(which could be due to defocus) and arranged in concentric manner with the smallest near the indenter contact point and expand outward during loading of the indenter due to change in stress and orientation of the sample which lead to a change in electron diffraction, some have dark contrast while others have bright contrast [8].

The system of cracks developed during the indentations consists of lateral cracks with variable lengths and are located under the contact surface with variable distance of up to ~100 nm. The median cracks with length >400 nm and width <10 nm are located at different positions in the residual deformation zone, some near the centre and some at distances of up to 500 nm from the centre. These median cracks run parallel to the direction of indentation and break the contact surface.

Figure 12a shows a high resolution image of a probable wear debris particle at the centre of the indentation site shown in Fig. 11b. The diffraction image (see Fig. 12b) of the same area shows extra spots due to probable wear debris particles in this location.

The in situ TEM results show the elastic behaviour of the alumina sample before fracture and also the process of crack development, Median cracks were observed to develop during the loading stage of the indenter which in agreement with earlier research findings [5]. These cracks break the surface in the area of the residual indentation site and their size depends on the load applied and the time of

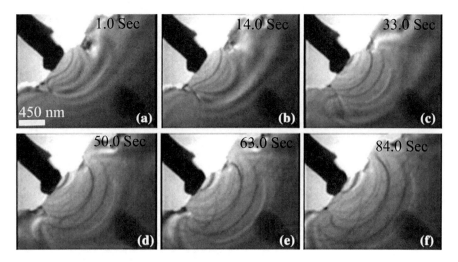

Fig. 9 a–f In-situ images of the nanoindentation of alumina with a 450 nm diameter blunt tungsten tip

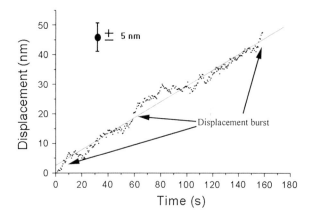

Fig. 10 Graph of the displacement of the indenter tip with time during deformation

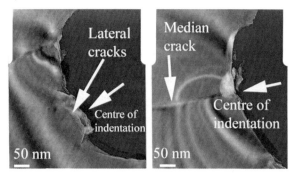

Fig. 11 TEM images of the damage area due to nanoindentation with a 450 nm diameter blunt tungsten tip, showing median and lateral cracks

Fig. 12 a High damage area due to nanoindentation, b Diffraction pattern from damage area

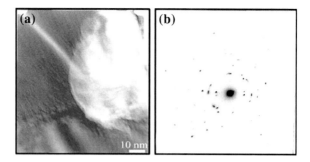

loading [5]. The median cracks were observed to develop from the contact surface and to extend downward during the loading time.

4 Conclusions

TEM analysis of the damage area beneath the microindentation sites in alumina reveals complex zones of defects, cracks and varying local strain.

- Cracks with variable sizes were observed to develop during the indentation process with material fragmentation.
- Median cracks were observed to develop in the deformation zone during loading and to propagate parallel to the indentation direction and their length >3 μm and width <10 nm.
- Lateral cracks were observed in the residual damage area near the surface, with variable size (some ~50 nm) and variable distance (up to 500 nm) from the centre of indentation.
- Bend contours, due to stress, were observed to develop prior to crack development during loading cycle of the indenter.
- Cracks with initial speed of 17 nm/s were observed to follow bend contours.

References

1. Wall, M.A., Dahmen, U.: An in situ nanoindentation specimen holder for a high voltage transmission electron microscope. Microsc. Res. Tech. **42**(4), 248–254 (1998)
2. Stach, E.A., Freeman, T., Minor, A.M., et al.: Development of a nanoindenter for in situ transmission electron microscopy. Microsc. Microanal. **7**(6), 507–517 (2001)
3. Bobji, M.S., Pethica, J.B., Inkson, B.J.: Indentation mechanics of Cu–Be quantified by an in situ transmission electron microscopy mechanical probe. J. Mater. Res. **20**(10), 2726–2732 (2005)
4. Bobji, M.S., Ramanujan, C.S., Pethica, J.B., et al.: A miniaturized TEM nanoindenter for studying material deformation in situ. Meas. Sci. Technol. **17**(6), 1324–1329 (2006)

5. Cook, R.F., Pharr, G.M.: Direct observation and analysis of indentation cracking in glasses and ceramics. J. Am. Ceram. Soc. **73**(4), 787–817 (1990)
6. Anton, R.J., Subhash, G.: Dynamic Vickers indentation of brittle materials. Wear **239**(1), 27–35 (2000)
7. Page, T.F., Oliver, W.C., McHargue, C.J.: The deformation behavior of ceramic crystals subjected to very low load (nano) indentations. Mater. Res. Soc. **7**(2), 450–473 (1992)
8. Chan, H.M., Lawn, B.R.: Indentation deformation and fracture of sapphire. J. Am. Ceram. Soc. **71**(1), 29–35 (1988)
9. Hockey, B.J.: Plastic deformation of aluminum oxide by indentation and abrasion. J. Am. Ceram. Soc. **54**(5), 223–231 (1971)
10. Tsui, T.Y., Vladdak, J., Nix, W.D.: Indentation plastic displacement field: part I the case of soft films on hard substrates. Mater. Res. **14**, 2196–2203 (1999)
11. Inkson, B.J., Bobji, M.S., Mobus, G., Kraft, O., Wagner, T.: TEM characterisation of nanoindentation deformation in Cu–Ti thin films sectioned by FIB. In: Proceedings of ICEM (2002)
12. Chaiwan, S., Hoffman, M., Munroe, P., et al.: Investigation of sub-surface damage during sliding wear of alumina using focused ion-beam milling. Wear **252**(7–8), 531–539 (2002)
13. Xie, Z.H., Munroe, P.R., Moon, R.J., et al.: Characterization of surface contact-induced fracture in ceramics using a focused ion beam miller. Wear **255**(1–6), 651–656 (2003)
14. Ma, L.W., Cairney, J.M., Hoffman, M.J., Munroe, P.R.: Characterization of TiN thin films subjected to nanoindentation using focused ion beam milling. Appl. Surf. Sci. **237**, 631–635 (2004)
15. Cairney, J.M., Harris, S.G., Munroe, P.R., et al.: Transmission electron microscopy of TiN and TiAlN thin films using specimens prepared by focused ion beam milling. Surf. Coat. Technol. **183**(2–3), 239–246 (2004)
16. Inkson, B.J., Leclere, D., Elfallagh, F., et al.: The effect of focused ion beam machining on residual stress and crack morphologies in alumina. J. Phys. Conf. Ser. **26**, 219–222 (2006)
17. Inkson, B.J., Steer, T., Mobus, G., et al.: Subsurface nanoindentation deformation of Cu-Al multilayers mapped in 3D by focused ion beam microscopy. J. Microsc. **201**(2), 256–269 (2001)
18. Steer, T.J., Mobus, G., Wagner, T., et al.: 3D FIB mapping of nanoindentation zones in a Cu-Ti multilayered coating. Thin Solid Films **413**(1–2), 147–154 (2002)
19. Inkson, B.J., Wu, H.Z., Steer, T.J., et al.: 3D mapping of subsurface cracks in alumina using FIB. In: Fundamentals of Nanoindentation and Nanotribology II. Material Research Society, Boston (2001)
20. Wu, H.Z., Roberts, S.G., Mobus, G., et al.: Subsurface damage analysis by TEM and 3D FIB crack mapping in alumina and alumina/5vol.%SiC nanocomposites. Acta Mater. **51**(1), 149–163 (2003)
21. Elfallagh, F., Inkson, B.J.: Evolution of residual stress and crack morphologies during 3D FIB tomographic analysis of alumina. J. Microsc. **230**(2), 240–251 (2008)
22. Molis, S.E., Clarke, D.R.: Measurement of stresses using fluorescence in an optical microprobe: stresses around indentations in a chromium-doped sapphire. J. Am. Ceram. Soc. **73**(11), 3189–3194 (1990)
23. Ostertag, C.P., Robins, L.H., Cook, L.P.: Cathodoluminescence measurement of strained alumina single crystals. J. Eur. Ceram. Soc. **7**(2), 109–116 (1991)
24. Ma, Q., Clarke, D.R.: Piezospectroscopic determination of residual stresses in polycrystalline alumina. J. Am. Ceram. Soc. **77**(2), 298–302 (1994)
25. Banini, G.K., Chaudhri, M.M., Smith, T., et al.: Measurement of residual stresses around Vickers indentations in ruby crystal using a Raman luminescence. J. Phys. D Appl. Phys. **34**, L122–L124 (2001)
26. Inkson, B.J.: Dislocations and twinning activated by the abrasion of Al_2O_3. Acta Mater. **48**(8), 1883–1895 (2000)

27. Farber, B.Y., Yoon S.Y., Lagerlof K.P.D., et al.: Microplasticity during high temperature indentation and the Peierls potential in sapphire (α-Al2O3). J. Phys. Stat. Sol. (a) **137**, 485 (1993)
28. Guillou, M.O., Henshall, J.L., Hooper, R.M.: Indentation fracture and soft impresser fatigue in sapphire and polycrystalline alumina. Int. J. Refra. Met. Hard. Mater. **16**(4–6), 323–329 (1998)
29. Farber, B.Y., Yoon, S.Y., Peter, K., et al.: Dislocation sources in sapphire (α-Al2O3) near microhardness indents. Mater. Res. Adv. Tech. **84**(6), 426–430 (1993)
30. Lockwood, A.J., Inkson, B.J.: In-situ TEM nanoindentation and deformation of Si-nanoparticle clusters. In: 14th European Microscopy Congress. Springer, Aachen (2008)
31. Stofel, E., Conrad, H.: Fracture and twinning in sapphire (Alpha-Al2O3) crystals. Trans. Metall. Soc. AIME. **227**(5), 1053 (1963)
32. Firestone, R.F., Heuer, A.H.: Creep deformation of 0° sapphire. J. Am. Ceram. Soc. **59**(1–2), 24–29 (1976)

The Effect of Nanotube Interaction on the Mechanical Behavior of Carbon Nanotube Filled Nanocomposites

Beril Akin and Halit S. Türkmen

Abstract This study focus on the effect of the mechanical interaction of the carbon nanotubes used as a filler material in a polymer composite. For this purpose, a representative volume element containing two-wavy carbon nanotubes is modeled by using the finite element method. A compressive displacement is applied to the representative volume element in the axial direction of the carbon nanotubes. Two type of analysis are done. These are the in-contact case in which a mechanical contact interaction between parallel adjacent faces of carbon nanotubes is defined and not defined later for analyzing the out-of-contact case. The effective modulus of elasticity is computed for each of them and the results are compared. The modulus of elasticity is found to be higher in the presence of contact between nanotubes.

Keywords Nanomechanics · Carbon nanotubes · Finite element methods · Contact mechanics

1 Introduction

Higher mechanical efficiency and lightness of composites materials make them indispensable for aerostructures. In this study we observed a more sophisticated composite material type; a nanocomposite consist of polymer matrix reinforced with carbon nanotubes.

B. Akin (✉) · H. S. Türkmen
Faculty of Aeronautics and Astronautics, Istanbul Technical University,
Maslak 34469 Istanbul, Turkey
e-mail: berilyumakn@gmail.com

H. S. Türkmen
e-mail: halit@itu.edu.tr

Using carbon nanotubes as a filler material increases the strength of the material and leads to a higher modulus of elasticity compared to the common composites. Carbon nanotube filled polymer nanocomposites are usually used for the strengthening poorly reinforced areas of the composite such as adhesive joint parts. There can be several parameters which affect the modulus of nanotube filled nanocomposites. The wavy shape of the nanotube such as studied by [1–4].

Two ways are appropriate to observe the mechanical behavior of nanocomposites. The molecular dynamics which focuses on atomic interactions and chemical reactions among the matrix and the filler material, the conventional continuum mechanical approach to observe the global behavior of the material, such as deformations, effective stiffness.

The representative volume element shape; square, cylindrical and hexagonal are studied by [5] to observe the carbon nanotube reinforcement effect having volume fractions of 2 and 5 % in a composite material. Their results have a good match with cited experimental studies in this field.

In this study, the contact between two nanotubes is modeled using the finite element method to observe the effect of nanotube interaction on the stiffness of the nanotube filled nanocomposites. The nanotube is considered as wavy hollow cylinder as mentioned in Luo et al. [6]. The nanotubes are modeled by using shell elements with a wall thickness of 0.34 nm coherently with Fisher et al. [4] and Liu et al. [7]. Carbon nanotube and polymer interface is assumed perfect. This idea is supported by Jia et al. [8] and Thostenson et al. [9] based on the assumption of a possible strong C–C bond at the interface (as cited in [5]).

The matrix surrounding the nanotubes is modeled using three dimensional solid elements. Only the quarter of representative volume is modeled in order to decrease the number of elements and consequently solution. The contact is defined between two adjacent nanotubes. A compressive displacement is applied in the axial direction to replicate the indentation test that is performed usually to obtain the modulus of elasticity experimentally. The displacement, strain and stress are obtained by performing the analysis. The analysis is also performed without defining the contact between two nanotubes. The modulus of elasticity values obtained from the analysis with and without contact are compared. The modulus of the elasticity is higher while two carbon nanotubes contact each other. Consequently, the interaction of carbon nanotubes have a stiffening effect for carbon nanotube filled nanocomposites.

2 Methods

2.1 Finite Element Model

The contact between nanotubes is modeled by using the finite element method due to following reasons. Firstly, the experimental studies of nanomaterials is usually seen as a chemistry subject. Secondly, contact interaction between carbon nanotubes get visible only by using a lot of sophisticated laboratory equipment. Thus, numerical modeling facilitate to predict nanoscale structures.

2.2 Modeling Considerations

The representative volume element is designed as a square geometry which contains two aligned curved carbon nanotubes modeled as hollow cylinders. The contact is defined between two adjacent nanotubes. The representative volume element is modeled only as a quarter for decreasing the number of elements and consequently solution time. A compressive displacement is applied in the axial direction to replicate the indentation test that is performed usually to obtain the modulus of elasticity experimentally. The analysis is also pursued without defining contact between two nanotubes to determining out of contact state.

The geometry can be assumed having a sinusoidal form as shown by Fischer et al. [4] the spatial position in the xyz Cartesian coordinates is given by

$$z = A \sin \frac{2\pi x}{\lambda} \quad (1)$$

The waviness ratio is A/λ A is the amplitude and λ is the wavelength of the function and length of the nanotube. In this study, carbon nanotube modeled have a waviness ratio of 0.2.

The representative volume element model (Fig. 1) is modeled via ANSYS®. The key points are generated via MATLAB® and implemented to ANSYS® via Ansys Parametric Design Language. The volume is created through a circle and a spline which is subtracted from a block of dimensions $400 \times 50 \times 250$ nm³. The matrix surrounding the nanotubes is modeled using three dimensional solid elements using element type 10-Node Tetrahedral Structural Solid 186. The meshes have triangular shape with pyramids having mid side nodes following curves. The model has 125384 element and 182091 nodes as can be seen in Table 1.

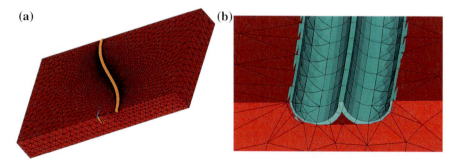

Fig. 1 The finite element model of the representative volume element (**a**). Larger view of the finite element model (**b**)

Table 1 Element and element numbers defined [10]

Label	Number	Name
Shell93	2880	8-Node structural shell
Solid186	119137	3-D 20-Node structural solid
Targe170	720	3-D target segment
Conta175	2647	Node-to-surface contact

The carbon nanotubes are meshed with element type 8-Node Shell 181. Element and meshes have quad shapes with pyramids having mid side nodes following curves. The analysis is solved by the pre-conjugated gradient PCG solver by four substeps, 34 iterations and 0.125 time increment.

Carbon nanotubes are considered isotropic having an elastic modulus E, 1000 GPa corresponded with (Kalamkarov 2006) [11] of single walled carbon nanotube having wall thickness of 0.34 nm and poisson ratio $v = 0.3$ similar in the studies of Goldstein et al. [12]. Generic material properties are considered for epoxy assumed as isotropic having elastic modulus $E = 5$ GPa and poisson ratio $v = 0.3$.

The finite element model is symmetrical on the $x = 0$ and $y = 0$ planes and displacement u_z is subjected to compression with a value different in each case studied.

Volume of epoxy model and carbon nanotube (considered straight) model are 4.99×10^6 nm^3 and 1570 nm^3, respectively. The Volume fraction of carbon nanotubes is obtained by considering the carbon nanotubes having straight forms.

$$V_f = \frac{V_{\text{CNTs}}}{V_{\text{CNTs}} + V_{\text{EPOXY}}} \quad (2)$$

The volume fraction $V_f = 0.000314$ is obtained for the representative volume element. In order to determine effective elastic modulus of the material rules of mixtures is used. Rules of mixtures also used by [5] to determine carbon nanotube reinforced nanocomposite which using in composite materials micromodeling. Furthermore, by the rules of mixtures,

$$E = V_f E_{\text{CNT}} + (1 - V_f) E_{\text{epoxy}} \quad (3)$$

$E = 5.312$ GPa is the effective modulus obtained without considering contact and waviness effects.

2.2.1 Contact Modeling

The contact is defined on the surfaces of carbon nanotubes. Intentionally, no material is modeled between the two carbon nanotube for a better observation. While the model is subjected to a compression stress, the model is dilating and carbon nanotubes are approaching and holding each other along parallel lines. The

Table 2 Contact parameters used [10]

Contact parameters	Factor or constant
Contact penalty stiffness	1 (factor)
Penetration tolerance	0.1 (factor)
Pinball region	3 (constant)
Contact normal	Normal to target surface
Contact algorithm	Augmented lagrange algorithm

friction is constant and defined as 0.1 respects with the static coefficient of friction values between graphite surfaces in a clean environment mentioned by Lide [13].

Due to the convergence difficulties the contact penalty stiffness is 1 as factor, penetration tolerance is 0.1, pinball region 3 as constant and contact normal is normal to target surface (Table 2).

3 Results

The objective is to obtain the displacement, strain and stress values by performing the analysis. The modulus of elasticity values are compared for analyzing in contact and out-of-contact states. The state of contact can be seen in Fig. 2 and the numerical values can be seen in Table 3. The numerical results are observed for the three cases where the model is subjected to a displacement of 20, 10 and 5 nm in compression.

Firstly, applying rules of mixtures to composite while considering carbon nanotubes as straight, modulus of elasticity is calculated. Then the numerical values are compared between the in-contact and out-of contact states. The effective elastic modulus is calculated via the well-know formulas from strength of materials,

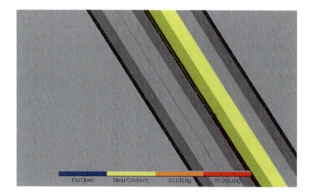

Fig. 2 Contact status

Table 3 Reaction forces and effective elastic modulus of representative volume element

Displacements		Representative volume element	Elastic modulus difference (%)
−20 nm	In-contact	$Rz = 8046$ nN $E = 5.0292$ GPa	0.307
	Out-of-contact	$Rz = 8021.4$ nN $E = 5.0138$ GPa	
−10 nm	In-contact	$Rz = 3919.6$ nN $E = 5.041$ GPa	0.059
	Out-of-contact	$Rz = 4010.8$ nN $E = 5.038$ GPa	
−5 nm	In-contact	$Rz = 2005.4$ nN $E = 5.0135$ GPa	0
	Out-of-contact	$Rz = 2005.4$ nN $E = 5.0135$ GPa	

$$E = \frac{P \cdot L}{\delta \cdot A} \quad (4)$$

where P is the total reaction force in the direction of the displacement, δ is the displacement, L is the length of the representative volume element without consideration of curvature effect and A is the cross sectional area.

The elastic modulus as predicted before, elastic modulus by the rules of mixtures with the filler volume fraction without considering contact and waviness effect is 5.12 GPa which is 1–2 % higher than the elastic modulus obtained by the finite element model.

The modulus enhancement compared to out-of-contact cases are 0.30 % for 20 nm and 0.059 % for 10 nm compression displacements. Apparently, strengthening is not proportional. Although, higher displacement produced more resistance and increased stiffness.

4 Discussion

From the results, it is obvious to see the contact contribution is increasing the effective elastic modulus of the model. A value of 0.3 % stiffness enhancement in the effective elastic modulus is observed. Consequently, the modulus of elasticity is higher while contact is occurring among two nanotubes. Consequently, the interaction of carbon nanotubes has a stiffening effect for carbon nanotube filled nanocomposites due to interaction effect. An experimental study for realizing a nanoindentation test can pursue this study to verify the results obtained.

References

1. Anumandla, V., Gibson, R.F.: A comprehensive closed form micromechanics model for estimating the elastic modulus of nanotube-reinforced composites. Compos. A Appl. Sci. Manuf. **37**(12), 2178–2185 (2006)
2. Bradshaw, R.D., Fisher, F.T., Brinson, L.C.: Fiber waviness in nanotube-reinforced polymer composites—II: modeling via numerical approximation of the dilute strain concentration tensor. Compos. Sci. Technol. **63**(11), 1705–1722 (2003)
3. Fisher, F.T., Bradshaw, R.D., Brinson, L.C.: Effects of nanotube waviness on the modulus of nanotube-reinforced polymers. Appl. Phys. Lett. **80**(24), 4647–4649 (2002)
4. Fisher, F.T., Bradshaw, R.D., Brinson, L.C.: Fiber waviness in nanotube-reinforced polymer composites—I: modulus predictions using effective nanotube properties. Compos. Sci. Technol. **63**(11), 1689–1703 (2003)
5. Liu, Y.J., Chen, X.L.: Evaluations of the effective material properties of carbon nanotube-based composites using a nanoscale representative volume element. Mech. Mater. **35**(1–2), 69–81 (2003)
6. Luo, D., Wang, W.-X., Takao, Y.: Effects of the distribution and geometry of carbon nanotubes on the macroscopic stiffness and microscopic stresses of nanocomposites. Compos. Sci. Technol. **67**(14), 2947–2958 (2007)
7. Liu, W.K., Wagner, G.J., Qian, D., Yu, M.F., Ruoff, R.S.: Mechanics of carbon nanotubes. Handbook of Nanoscience, Engineering, and Technology (Chapter 19). CRC Press, Boca Raton (2002)
8. Jia, Z., Wang, Z., Xu, C., Liang, J., Wei, B., Wu, D., Zhu, S.: Study on poly(methyl methacrylate)/carbon nanotube composites. Mater. Sci. Eng. A **271**(1–2), 395–400 (1999)
9. Thostenson, E.T., Ren, Z.F., Chou, T.-W.: Advances in the science and technology of carbon nanotubes and their composites: a review. Compos. Sci. Technol. **61**, 1899–1912 (2001)
10. ANSYS® Academic Research, Analysis Guide. Accessed June 2012
11. Kalamkarov, A.L.: Analytical and numerical techniques to predict carbon nanotubes properties. Int. J. Solids. Struct. **43**(22–23), 6832–6854 (2006)
12. Goldstein, R.V., Gorodtsov, V.A., Chentsov, A.V., Starikov, S.V.: Description of mechanical properties of carbon nanotubes. Size effect. Part 2. Письма о материалах т.1, 190–193 (2011)
13. Lide, D.R.: Handbook of Chemistry and Physics. CRC Press, Boca Raton, pp. 15–40 (1994)

An Automatic Process to Identify Features on Boreholes Data by Image Processing Techniques

Fabiana Rodrigues Leta, Esteban Clua, Mauro Biondi, Toni Pacheco and Maria do Socorro de Souza

Abstract The breakouts and drilling-induced tensile fractures in borehole walls give a good estimation of the in situ stress orientation. While it is manually simple to analyze images of this nature, this task may become tedious or even impossible when a large amount of data is present. In this work we propose a novel image based analysis process to automatically guide the geologist when using images. For the interpretation of images we considered the position of the roller tool to see if the events were produced or not by the drag tool. While the analysis may be not as accurate as made with human vision, our proposed method may be used as a pre-processing stage, separating a set of regions that will be carefully treat in a manually process.

Keywords Image segmentation · Oil well visualization · Image log · Closed natural cubic splines

---

F. R. Leta (✉) · M. Biondi · M. do Socorro de Souza
Computational and Dimensional Metrology Laboratory, Mechanical Engineering Department, Federal University Fluminense -UFF, Niterói, RJ 24210-240, Brazil
e-mail: fabianaleta@id.uff.br

M. Biondi
e-mail: mbiondi@lmdc.uff.br

M. do Socorro de Souza
e-mail: souza.msocorro@gmail.com

E. Clua · T. Pacheco
Computing Institute, Federal University Fluminense -UFF, Niterói, RJ 24210-240, Brazil
e-mail: esteban@ic.uff.br

T. Pacheco
e-mail: tpacheco@lmdc.uff.br

1 Introduction

It is possible to minimize costs and risks in the process of drilling boreholes in a determined area with the analysis of its soil structure. But even for geologists, the study of oil and gas drilling locals can be difficult because it relies on the visualization and interpretation of the data obtained [1].

According to Aadnoy [2], for years people have estimated the time loss associated with unexpected borehole stability problems to account for 10–15 % of the time required to drill a well. Since the rig time is the major cost factor in drilling operation, we understand that borehole stability problems are very costly for the industry. Borehole collapse is possibly the most costly single problem encountered during drilling of a well, and there is not a trivial solution for the problem.

This theory states that drilling generates changes in the stress field of the formation due to supporting material losses, inducing stresses that can result in more trouble [3]. If the stress is higher than the rock strength, rocks can cause the borehole collapse. So the mainly fundamental of the solution proposed in this work is the possibility to prevent and reduce instability problems.

The better the planning of a well, the greater are the chances of success to be achieved. In this context, the word success means achieving the objectives of the project according to the safety standards of the company, time and costs consistent with the ones of the market. However, to achieve this success, geologists, reservoir engineers and production engineers should not only care to the optimization of the production of the reservoir but should also be aware of the operational risks and expenses on the project.

When the well is drilled and a part of its formation is removed, there is instability of stresses. In that moment comes one of the most important stages of the project of a well which is ensuring the stability of it, and to achieve this goal it is necessary to evaluate factors like: rock resistance, temperature variations, well trajectory and geometry.

The breakouts and drilling-induced tensile fractures in borehole walls give a good estimation of the in situ stress orientation. In vertical wells or near-vertical wells, the breakout axis directly indicates the orientation of the minimum horizontal stress (h), if the mud weight is too low, while drilling-induced tensile fracture is oriented in the direction of the maximum horizontal stress (H), if the mud weight is high. Borehole breakouts and drilling-induced tensile fracture occur in pair on opposite sides of the borehole. The vertical tensile fracture occurs at exactly 90° to the orientation of the breakouts, but generally, not at the same depth as the breakouts.

Typically the image analysis is manually made and used to identify natural features or induced features by drilling on the walls of a borehole. These images are obtained from the geophysical logging tools such as the ultrasonic borehole televiewer (BHTV), micro acoustic (UBI, CBIL) or micro resistivity (FMI, OBMI, EMI) and the caliper logs.

We can identify features, natural or not, displayed on these images. Examples of this features are: natural fractures, beddings, vugs, drilling-induced tensile fractures, and breakouts.

In this work we propose a novel image based analysis process for automatically guide the geologist when using images. For the interpretation to images we considered the position of the roller tool to see if the events were produced or not by the drag tool.

We finally present some results obtained from the analysis of these images, with the aim of supporting the user in analyzing and planning the drilling.

2 Related Works

No well is drilled without problems [4]. Managing drilling risks means not letting small problems become big ones. Knowing what the risks are and when they are likely to occur, keeps surprises to a minimum possibility. Most of the time and cost spent in drilling is encountered not in the reservoir, but in its planning. For instance drilling at a high rate of penetration can save time and money. However, when accompanied by a too low drill string rotation rate or mud flow rate, which fails to lift rock cuttings to surface, the result is a stuck pipe. Faults and fractures encountered by the wellbore open conduits for loss of drilling fluid to the formation. Excessive high mud pressure can fracture the formation and cause lost circulation. Too low mud pressure fails to keep high-pressure formations under control, resulting in gas kicks or worse, blowouts.

Jarosinski and Zoback present a borehole visualization method focused on stress and breakout analysis [5]. In [6] and [7], the authors present important factors to be considered in order to model the requirements associated to a detailed data processing and analysis of this kind of data, that will be considered on the proposed system.

Luthi and Souhait [8] detected and traced features, generically termed "fractures", on Formation MicroScanner images using simple statistical techniques which take into account natural variations of the rock matrix conductivity . Forward modeling results for the Formation MicroScanner response in front of a fracture are incorporated into an inversion scheme to extract fracture traces and to compute their fracture apertures. Visual comparison of the resulting fracture traces with the original Formation MicroScanner images is good, even for complex intersecting fracture networks.

Tilke et al. [9] demonstrated that geostatistical analysis can be applied to borehole image data in order to generate an improved understanding of formation heterogeneity, looking at the pore/vugs distributions on images. They present a technique to quantitatively describe the porosity heterogeneity in a borehole at the scale of several tenths of an inch.

In [1] the authors present a technique for modeling a 2D approach, so important requirements of these structures are maintained and can be analyzed, such as perforation fluid weight, rock strength and the hole geometry. Figure 1 presents the data obtained by a four arms caliper log. The 2D developed system shows an approximation of the borehole shape in a specific height (Fig. 2). Each colored profile is related to a caliper arm and is used for the 3D reconstruction.

The obtained data can be three-dimensional modeled to help the borehole visualization [10]. The modeling is based on splines [11, 12].

Sensors installed at the Caliper System also achieve images with soil characteristics. These images may also contain details that characterize breakouts and other features. Figure 3 shows some examples obtained by this tool.

3 Image Processing

For the identification of the main events obtained by the caliper tool, it is necessary to enhance them using some image processing algorithms.

For the edge enhancement and the extraction of its points coordinates, there are many different approaches. We can highlight the following edge enhancement algorithms: Sobel, Prewitt, Roberts, Laplacian, Zero-Crossing and Canny [13] [14]. These digital filters have been used in many different applications, as in [15] that present an edge enhancement algorithms comparison to be used in the measurement of standards aperture area. It is important to apply a filter before any segmentation step, since the obtained images from Caliper are suitable to many noise entrances.

For this work we propose the usage of a simple filter, called the median filter. The basic idea consists on visiting each entrance value and order all the horizontal and vertical neighborhoods, substituting the value itself by the middle value of the

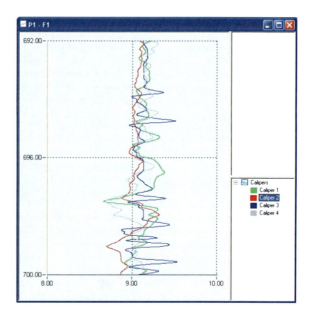

Fig. 1 Profiles obtained by a 4-arm caliper log

An Automatic Process to Identify Features

Fig. 2 A shot of the 2D caliper analysis system

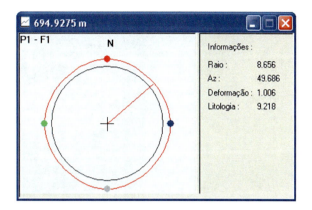

ordered sequence. For border values, the algorithms suggest to fill with the same border value. For a simple example, notice the following one-dimensional case:

Data = [... 3 6 4 ...]

Filtered_Data = [... 3 4 4 ...]

Once the main features are enhanced the next step is to encode these features in the image. The use of a discrete boundary code, such as "chain codes" can be an adequate solution. Considering a set of discrete boundary orientation, it is possible to obtain a polygonal boundary of a contour segment. We can specify, for instance, a set of 8 different codes, representing orientation displacements, like: horizontal up, horizontal down, vertical right, vertical left, diagonal right and up, diagonal right and down, etc. Then the feature contour can be represented in terms of a sequence of codes (numbers). It is a syntactic description of the boundary conditions [16].

Although there are many different chain codes, in the case of geometric features, which can be coded as equations, the Hough Transformation becomes a better solution. In Fig. 3 the features have similar aspects with the sinusoids, so it is possible to use the Hough Transformation, which is able to detach geometric forms in binary images from its parametric equations. The Hough Transformation is a well-established and common method for computer vision problems [16–18].

According to Schalkoff [16], Hough techniques are particularly useful for computation of sets of global description parameters (perhaps noisy) from local measurement.

The method consists on applying the transformation to the image in such a way that all the pixels that are in the same curve can be mapped through one single point of a new parameterized space of the searched curve.

The idea of the Hough transform is to transform the image in the digital space as a representation of the parameters described by the type of curve that is desired to be found in the image. This transformation is applied so that all points belonging to the same curve are mapped into a single point in the parameter space of the curve searched. For this, the parameter space is discretized and represented as an array of integers, where each position of the array, which satisfies the searched curve equation, increments the counter corresponding to its position in the

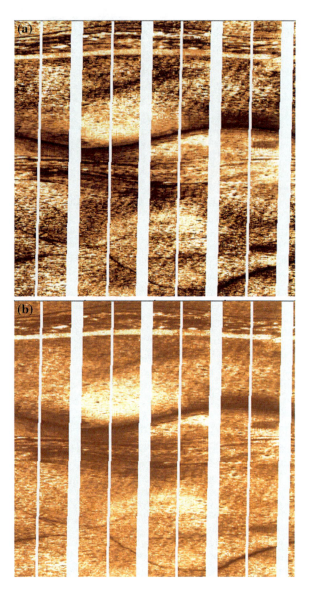

Fig. 3 Samples obtained by micro resistive tool: **a** dynamic image and **b** static image

discretized representation (matrix). In the end, the accountant who has the highest value will correspond to the parameters of the curve described in the image [14].

In Fig. 4 we present an example of possible curves in one point in the image, considering the Hough Transformation.

Depending on the geometric form, parameters must be defined previously. For instance, with the Hough transformation used for identifying circumferences in images it is fixed a circle radius, while at the sinusoid transformation it is necessary to fix as parameters the curve amplitude.

Fig. 4 Possible curves in one point in the image

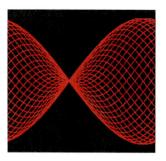

4 Methodology

For this work, we considered two groups of images obtained from the Caliper tool: dynamic and static. Each one is generated with different hardware strategies. The purpose of this approach consists on identifying events with sinusoidal aspects from the borehole region, in order to be able to analyze in future works events that may turn these regions compromised.

For each analyzed image we developed the following pipeline:

1. Image acquisition—obtained from Caliper and transformed to color images, with a customized colorization process;
2. Pre-processing—This stage involves the application of techniques that allow reducing the computational complexity and facilitating the image processing in the following stage. The pre-processing was sub-divided into the following steps:
    a. Transformation of the colorized image (with pseudo-color information) into a gray scale image, with 256 gray values;
    b. Quantification: the image is quantified into 32 gray levels in order to turn the process easier and faster;
    c. Application of the medium filter, in order to simplify and reduce the noise of the image; This is done with a 9 × 9 filter mask;
    d. Image segmentation: we choose a threshold of the image histogram, considering the cut value of the color, corresponding to 3 % of the total amount of pixels.
    e. Transform the image into its negative one;
3. Processing: At this stage we identify the events with sinusoidal formats, using the Hough transformation;
4. Result presentation: The obtained results of the processed images are transferred to the original result and used for the context analysis.

In our proposal, due the specialized aspect of the problem, we started choosing filtered points obtained from the detected curves, establishing sizes from 10 to 120 pixels. For each size, it is calculated a particular Hough transformation and it is established a punctuation value for the specified curve. After the iterations, our algorithm chooses the curve that received more points, considering it as the best candidate.

The complexity of this algorithm is related to the fact that each point is added to the processing stage and 360 possible curves are selected as candidates. There are then 360 × N incremental operations at the punctuation map, being N the total number of points. The searching of the most punctuated point is a linear operation that goes once through the entire map. This algorithm is executed only one time for each fixed image size, resulting in a M (amplitude) × N (points) size.

The execution time varies depending on the analyzed image, considering how efficient the pre-processing stage is. This means that if the process of the Hough transformation detaches only relevant points, eliminating noises and not valid regions, the sinusoidal curve detection will be more efficient, in relation with the analysis point reduction.

5 Analysis and Information

The first experimental case used in this work was based on a acoustic dynamic image with borehole events that may be representing fractures (Fig. 5). It is possible to see that after the image pre-processing the Hough transformation clearly identifies one of the fractures at the upper part of the image.

In order to detach other possible fractures and events, it is necessary at this stage of the work to proceed with the searching process looking other most ranked curves in the remaining of the image, being trivially possible to identify other details that are in different regions (Fig. 6).

The second test of this work used an acoustic image with less noise than the first and a clear fracture event, facilitating its identification with the Hough transformation, after the pre-processing (Fig. 7).

For the following case (Fig. 8), we used the same methodology that was applied for the static images, mentioned in the first case. The obtained results are similar to the dynamic images previously analyzed. It can be noted that the results are not so suitable for the static case. This is due to a difference of observed contrasts. In the case of the dynamic image (FMI), where there is a better contrast, the results were more accurate, while at the static image, with low contrast, it is possible to see a blur region with dark points, which causes a high rank punctuation when the Hough transformation is applied. This causes a highlight in a region where there is no event or fracture.

An Automatic Process to Identify Features

Fig. 5 Case 1: **a** original;
b *gray scale*; **c** 32 levels
quantized; **d** median filter;
e threshold; **f** negative;
g Hough transformation;
h Hough transformation
inserted in the original image

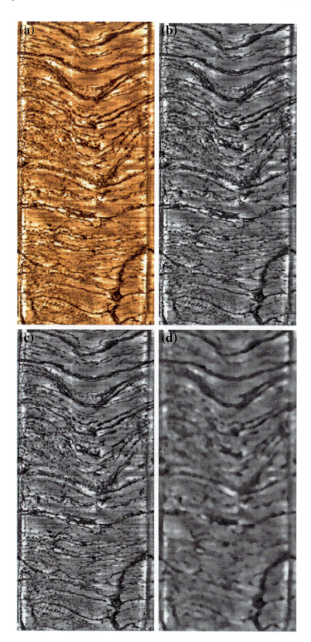

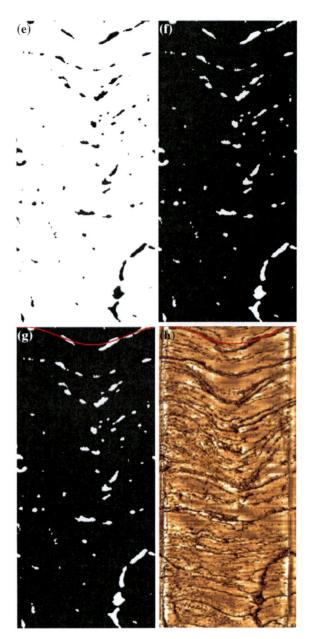

Fig. 6 Case 1: (e) threshold; (f) negative; (g) Hough transformation; (h) Hough transformation inserted in the original image

An Automatic Process to Identify Features 259

Fig. 7 Case 2: **a** *gray scale*; **b** negative, after quantization and median filter; **c** Hough transformation; **d** Hough transformation inserted in the original image

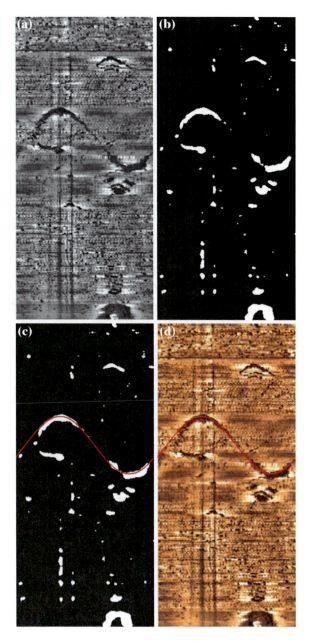

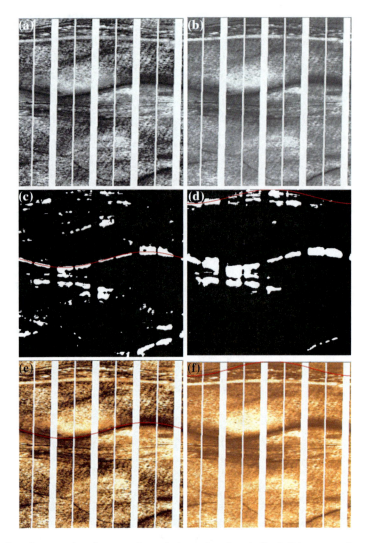

Fig. 8 Case 3: comparison between dynamic (**a, c, e**) and static (**b, d, f**) images: **a, b** gray scale images; **c, d** Hough transformation; **e, f** Hough transformation inserted in the original image

6 Conclusions

In this paper the authors presented a novel methodology based on traditional image segmentation for automatic and precise features extractions of caliper based images, typically from oil bare holes.

While the accuracy of the information is not comparable as when manually analyzed, our method showed to be an excellent choice for a first pre-processing analysis of a large amount of images. Results show that for more precise event extractions, it is necessary to manually change parameters.

Other segmentation methods may be included in this work and compared with ours. It is also possible to include machine learning [19], in order to feed the segmentation process with more accurate data. While the images may contain many different aspects, this machine learning approaches may include vector patterns that can easily be enhanced.

Our method was executed in sequential process, but may be easily paralyzed. With the usage of GPUs and multiple threads architectures, it should be possible to implement an interactive process, in order to give immediately feedback to the user.

The obtained results show that the methodology is feasible, but it is necessary to improve it in order to identify other events, not only the main event in the image. The use of Hough Transformation, considering a parametric sinusoid, is satisfactory to highlight the most important events in a borehole, making possible the study of the borehole tensile state more accurately.

Acknowledgments The authors would like to acknowledge Petrobras Oil and FAPERJ (E-26/171.362/2001) for the financial supported.

References

1. Leta, F.R., Souza, M., Clua, E., Biondi, M., Pacheco, T.: Computational system to help the stress analysis around boreholes in petroleum industry. In: Proceedings of the ECCOMAS 2008, Venice (2008)
2. Aadnoy, B.S.: Modern Well Design. A.A. Balkema, Rotterdam (1996). ISBN 90 54106336
3. Jiménez, J.C., Lara, L.V., Rueda, A., Trujillo, N.S.: Geomechanical wellbore stability modelling of exploratory wells—study case at Middle Magdalena Basin. C.T.F Cienc. Tecnol. Futuro **3**(3), 85–102 (2007). ISSN 0122-5383
4. Aldred, W., Plumb, D., Bradford, I., Cook, J., Gholkar, V., Cousins, L., Minton, R., Fuller, J., Goraya, S., Tucker, D.: Managing Drilling Risk. Oilfield Review. Summer (1999)
5. Jarosinski, M., Zoback, M.D.: Comparison of six-arm caliper and borehole televiewer data for detection of stress induced wellbore breakouts: application to six wheals in the Polish Carpathians, pp. F8-1–F8-23 (1998)
6. Barton, C., Zoback, M.D.: Stress perturbations associated with active faults penetrated by boreholes: possible evidence for near-complete stress drop and a new technique for stress magnitude measurements. J. Geophys. Res. **99**, 9373–9390 (1994)
7. Peska, P., Zoback, M.D.: Compressive and tensile failure of inclined well-bores and direct determination of in situ stress and rock strength, J. Geophys. Res. **100**, 12791–12811 (1995)
8. Luthi, S.M., Souhait, P.: Fractures apertures from electrical borehole scans. Geophysics **55**(7), 821–833 (1990)
9. Tilke, P.G., Allen, D., Gyllensten, A.: Quantative analysis of porosity heterogeneity: application of geostatistics to borehole images. Math. Geol. **38**(2), 155–174 (2006)
10. Barboza, D., Gazolla, J., Biondi, M., Clua, E., Leta, F. Tridimensional geometry reconstruction of oil and gas boreholes. In: ACE-X (2010)
11. Reinsch, C.: Smoothing by spline functions. Numer. Math. **10**, 177–183. Proceedings of the XII SIBGRAPI (October 1999), pp. 101–104 (1999)

12. Lambert, T.: Closed Natural Cubic Splines. Software Available from: www.cse.unsw.edu.au/~lambert/splines/natcubicclosed.html. Accessed May 01 2013
13. Duda, R., Hart, P.: Use of the Hough transformation to detect lines and curves in pictures. Commun. ACM **15**(1), 11–15 (1972)
14. Conci, A., Azevedo, E., Leta, F.R.: Computação Gráfica – Teoria e Prática, [v.2]. Elsevier Editora Ltda, Rio de Janeiro (2008)
15. Costa, P.B., Leta, F.R.: Measurement of the aperture area: an edge enhancement algorithms comparison. In: IWSSIP 2010—17th international conference on systems, signals and image processing, Rio de Janeiro (2010)
16. Schalkoff, R.J.: Digital Image Processing and Computer Vision, p. 489. Wiley, New York (1989)
17. Hough, P.V.C.: Machine Analysis of Bubble Chamber Pictures. In: Proceedings of the international conference high energy accelerators and instrumentation (1959)
18. Bovik, A.: Handbook of Image and Video, 2nd edn. Elsevier Academic Press, Burlington (2005)
19. Lattner, A.D., Miene, A., Herzog, O. A.: Combination of machine learning and image processing technologies for the classification of image regions. In: Proceeding of: adaptive multimedia retrieval: first international workshop, AMR, Hamburg, Germany, September, pp. 15–16 (2003)

An Optimization Procedure to Estimate the Permittivity of Ferrite-Polymer Composite

Ramadan Al-Habashi and Zulkifly Abbas

Abstract A numerical optimization method is performed using the MATLAB program to estimate the relative complex permittivity of each component of Samarium-substituted Yttrium Iron Garnet nanoparticles in Poly-vinylidenefluride (Sm-YIG-PVDF) composite samples. The optimization is taken as the optimized parameters that yield a minimum sum for the absolute differences between the calculated impedance obtained by using the permittivity calculated from Maxwell–Garnett (MG) formula and the measured equivalent one over the entire frequency range named the objective function (M). The guessed (estimated) ranges of the complex permittivity are based on the measured values of each component of Sm-YIG-PVDF composite samples. The optimized (optimum) impedance values are in very good agreement with the measured one for each composite and within the estimated ranges. More details on the optimization procedure are illustrated, and the permittivity of different composition dependence on the mole fraction of the Sm-YIG-PVDF composite materials is shown.

Keywords MATLAB · Permittivity · Sm-YIG-PVDF composite · Maxwell–Garnett

---

R. Al-Habashi (✉)
Physics Department, Azzaytuna University, Tripoli, Libya
e-mail: rtasneem2000@gmail.com

Z. Abbas
Faculty of Science, Physics Department, Advanced Materials and Nanotechnology Laboratory (AMNL), Institute of Advanced Technology (ITMA), Universiti Putra Malaysia, 43400 UPM Serdang, Selangor, Malaysia
e-mail: za@science.upm.edu.my

1 Introduction

Ferrite-polymer composites have extensive potential for applications in microwave devices and electromagnetic wave absorbers [1–10]. The ability to design the desired properties of the composite material could trigger new variety of applications.

The best known mixing rule is the Maxwell–Garnett (MG) formula for the effective complex permittivity of two phases mixture material [11]. Study and optimize the electromagnetic properties of the composites materials via numerical techniques or improving these techniques to be able to design the materials for specific application, will save time and costs, instead of designing composite materials with classical trial and error routine. This study was carried out to estimate the permittivity of ferrite-polymer composite with a new optimization procedure, in order to have good agreement results with the measured permittivity values of the composite.

2 Methodology

RF (Radio Frequency) Impedance/Material Analyzer (Agilent 4291B, 1 MHz to 1.8 GHz) was used to measure the complex relative permittivity of the composite material [12]. The analyzer calculated the relative permittivity of Sm-YIG-PVDF composite samples from their measured admittance according to the Eq. (1):

$$\varepsilon = \frac{Y_m}{j\omega c_o} \quad (1)$$

where, $Y_m = (1/Z_m)$ is the measurement admittance value of the material under test (MUT). C_o is the capacitance value of the air gap (whose distance between the analyzer's electrodes is same as the thickness of the MUT) [12].

The numerical optimization was performed using the MATLAB program [13] to estimate the relative complex permittivity of each component of the Sm-YIG-PVDF composite material in the frequency range of 10 MHz to 1 GHz (Fig. 1).

The criterion for optimization was taken to be that the optimized parameters yield a minimum sum for the absolute differences between the calculated impedances obtained by using the permittivity calculated from MG formula Eq. (2), and measured equivalent ones over the entire frequency range named the objective function (M) Eq. (3), as follow:

$$k = k_1 - 3k_1 \left(\frac{f\left(\frac{k_2 - k_1}{k_2 + 2k_1}\right)}{1 - f\left(\frac{k_2 - k_1}{k_2 + 2k_1}\right)} \right) \quad (2)$$

An Optimization Procedure

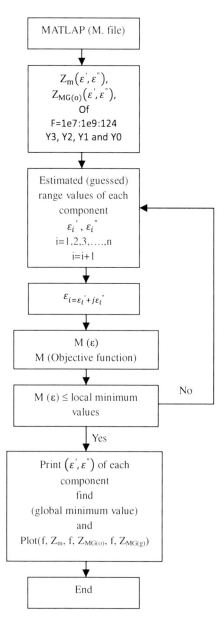

Fig. 1 Chart of the optimization process using the MATLAB program

where, k_1 is the permittivity of the host material, and that for the guest is k_2. The important parameter is the volume fraction of the inclusions f. The volume fraction occupied by the host is $1-f$.

$$M = \sum_{i=1}^{n} \left(\frac{|Z_m^2| - |Z_{MG}^2|}{n} \right) \qquad (3)$$

where, Z_m and Z_{MG} are the measured and the calculated impedance by using the permittivity obtained from the MG formula, respectively $n = 124$, is the number of measured or calculated impedance data points of the composite. This is to estimate the effective complex permittivity of each component of a composite material. The guessed or estimated ranges of the complex permittivity are based on the measured values of each component of the Sm-YIG-PVDF composite samples.

The MATLAB program will repeat the calculation with new estimated values of the permittivity of the material of interest via a loop to find the local minimum values of the objective function. This loop repeats the calculations a specified number of times until the objective function condition is satisfied at the global minimum value.

The global minimum value is given at the lower minimum value of the local minimum values of the objective function M. When the objective function condition is satisfied, the program gives the estimated value of the complex permittivity of the investigated composite sample [13–16].

3 Results

The effective complex permittivity calculated via MG formula with estimated (guessed) values of each component by using a MATLAB program based on a designed objective function to minimize the difference between the calculated and measured impedance values of Sm-YIG-PVDF composites. However, the guessed or estimated ranges of the complex permittivity are based on the measured values of each component of Sm-YIG-PVDF composite samples.

Measured impedance, calculated and optimized one based on MG formula of Y3 ($Y_3Fe_5O_{12}$) in PVDF composite are presented in Fig. 2. The estimated ranges of the real and imaginary complex permittivity of the Y3 are from 4.5 to 10.5, and

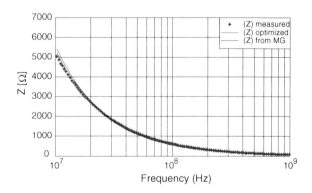

Fig. 2 Optimized (optimum values), calculated MG and measured impedance vs frequency of Y3 in PVDF composite

An Optimization Procedure

Table 1 Estimated and optimized values of relative permittivity and, objective function of Y3 in PVDF composite

Estimated range	Optimized values	Objective function value		
ε_{ry}	ε_{rp}	ε_{ry}	ε_{rp}	M
$\varepsilon_{r'}$ (4.5–10.5) and $\varepsilon_{r''}$ (0.07–0.13)	$\varepsilon_{r'}$ (1.5–5.5) and $\varepsilon_{r''}$ (0.3–0.7)	9.47 − j 0.1197	3.63 − j 0.513	13.2926
		8.83 − j 0.1133	3.64 − j 0.514	0.8426
		7.69 − j 0.1019	3.66 − j 0.516	25.6397
		5.85 − j 0.0835	*3.70 − j 0.520*	*0.6611*
		5.46 − j 0.0796	3.71 − j 0.521	12.1518

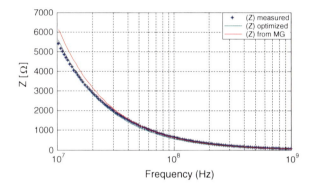

Fig. 3 Optimized (optimum values), calculated MG and measured impedance vs frequency of Y2 in PVDF composite

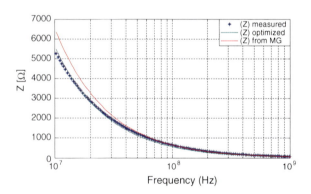

Fig. 4 Optimized (optimum values), calculated MG and measured impedance vs frequency of Y1 in PVDF composite

from 0.07 to 0.13, respectively. The guessed ranges of the real and imaginary complex permittivity of the PVDF are from 1.5 to 5.5, and from 0.3 to 0.7, respectively (Table 1). It can be observed that, the optimized (optimum) impedance is a very close to the measured one. This is indicated that, the optimization process eliminated the difference between the measured impedance and the calculated one by MG formula via a specific objective function M.

The optimum estimated values of the effective complex permittivity within the specified limits estimated range of each complex permittivity component of Y3 in

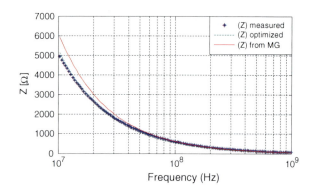

Fig. 5 Optimized (optimum values), calculated MG and measured impedance vs frequency of Y0 in PVDF composite

Table 2 Estimated and optimized values of relative permittivity, and objective function of Y2 in PVDF composite

Estimated range	Optimized values	Objective function value		
ε_{ry}	ε_{rp}	ε_{ry}	ε_{rp}	M
$\varepsilon_{r'}$ (6.5–10.5) and $\varepsilon_{r''}$ (0.1–0.5)	$\varepsilon_{r'}$ (1.5–5.5) and $\varepsilon_{r''}$ (0.3–0.7)	10.195−j 0.4695	3.9200−j 0.542	4.6745
		8.962−j 0.3462	3.9400−j 0.544	1.9562
		8.409−j 0.2909	3.9500−j 0.545	8.8352
		7.411−j 0.1911	*3.9700−j 0.547*	*0.2969*
		6.533−j 0.1033	3.9900−j 0.549	4.8368

Table 3 Estimated and optimized values of relative permittivity and, objective function of Y1 in PVDF composite

Estimated range	Optimized values	Objective function value		
ε_{ry}	ε_{rp}	ε_{ry}	ε_{rp}	M
$\varepsilon_{r'}$ (10.5–16.5) and $\varepsilon_{r''}$ (2.5–8.5)	$\varepsilon_{r'}$ (1.5–5.5) and $\varepsilon_{r''}$ (0.3–0.7)	16.012−j 8.012	4.20−j 0.570	0.8513
		14.398−j 6.398	4.22−j 0.572	3.9931
		12.372−j 4.372	4.25−j 0.575	3.4781
		11.775−j 3.775	4.26−j 0.576	2.2664
		11.209−j 3.209	4.27−j 0.577	3.8249

PVDF composite calculated at the global minimum objective function value 0.6611 (Table 1). All the optimized values of the complex permittivity for both components [$Y_3Fe_5O_{12}$ (Y3) and PVDF] are within the estimated ranges which presented in Table 1.

This optimization procedure was applied for the rest of the Sm-YIG in PVDF composite samples [Y2 ($Y_2Sm_1Fe_5O_{12}$), Y1 ($Y_1Sm_2Fe_5O_{12}$) and Y0 ($Sm_3Fe_5O_{12}$)]. Their measured impedances, calculated and optimized based on MG formula to estimate the relative complex permittivities are presented in the Figs. 3, 4, 5. However, the estimated range, optimized complex permittivities for both components per each composite and the objective function for each sample are presented in Tables 2, 3 and 4.

Table 4 Estimated and optimized values of relative permittivity, and objective function of Y0 in PVDF composite

Estimated range	Optimized values	Objective function value		
ε_{ry}	ε_{rp}	ε_{ry}	ε_{rp}	M
$\varepsilon_{r'}$ (7.0–12.0) and $\varepsilon_{r''}$ (0.1–0.6)	$\varepsilon_{r'}$ (1.5–5.5) and $\varepsilon_{r''}$ (0.3–0.7)	11.093−j 0.5093	4.22−j 0.572	4.3078
		10.443−j 0.4443	4.23−j 0.573	0.7030
		9.271−j 0.3271	4.25−j 0.575	0.3856
		8.242−j 0.2242	4.27−j 0.577	1.6228
		7.773−j 0.1773	4.28−j 0.578	1.4355

4 Conclusion

A numerical optimization method was performed using the MATLAB program to estimate the relative complex permittivity of each component of ferrite-polymer composites. The optimum estimated values of the relative complex permittivity within the specified limits estimated range of each complex permittivity component of Sm-YIG in PVDF composites calculated at the global minimum objective function value.

It was found that the optimized (optimum) impedance is very close to the measured one of each composite, and the optimized values of the complex permittivity for both components [Sm-YIG and PVDF] are within the estimated ranges. This indicates that the optimization process eliminated the difference between the measured impedance and the calculated one by MG formula via a specific objective function.

Acknowledgments Azzaytuna University-Libya, Nano and Advanced Technology Project under National Authority of Scientific Research (NASR), Tripoli, Libya and Universiti Putra Malaysia (UPM) are greatly acknowledged.

References

1. Salamone, J.C.: Polymeric Materials Encyclopedia, vol. 4, p. 2536. CRC Press, Boca Raton (1996)
2. Salamone, J.C.: Polymeric Materials Encyclopedia, vol. 9, pp. 7115–7126. CRC Press, Boca Raton (1996)
3. Chen, P., Wu, R.X., Zhao, T., Yang, F., Xiao, J.Q.: Complex permittivity and permeability of metallic magnetic granular composites at microwave frequencies. J. Phys. D. Appl. Phys **38**, 2302–2305 (2005)
4. Gupta, N., Kashyap, S.C., Dube, D.C.: Microwave behavior of substituted lithium ferrite composites in X-band. J. Magn. Magn. Mater. **288**, 307–314 (2005)
5. Kimura, S., Kato, T., Hyodo, T., Shimizu, Y., Egashira, M.: Electromagnetic wave absorption properties of carbonyl iron-ferrite/PMMA composites fabricated by hybridization method. J. Magn. Magn. Mater. **312**(1), 181–186 (2007)
6. Yang, Q., Zhang, H., Liu, Y., Wen, Q., Jia, L.: The magnetic and dielectric properties of microwave sintered yttrium iron garnet (YIG). J. Mater. Lett. **62**(17–18), 2647–2650 (2008)

7. Abbas, S.M., Dixit, A.K., Chatterjee, R., Goel, T.C.: Complex permittivity, complex permeability and microwave absorption properties of ferrite-polymer composites. J. Magn. Magn. Mater. **309**(1), 20–24 (2007)
8. Dosoudil, R., Ušáková, M., Franek, J., Sláma, J., Olah, V.: RF electromagnetic wave absorbing properties of ferrite polymer composite materials. J. Magn. Magn. Mater. **304**(2), e755–e757 (2006)
9. Abbas, Z., Pollard, R.D., Kelsall, R.W.: Complex permittivity measurements at Ka-band using rectangular dielectric waveguide. IEEE Trans. Instrum. Meas. **50**(5), 1334–1342 (2001)
10. Chen, L., Varadan, V.V., Ong, C.K., Neo, C.P.: Microwave electronics: measurement and materials characterization. Wiley, New Jersey (2004)
11. Priou, A.: Dielectric properties of heterogeneous materials. Elsevier Science Publishing Co., Inc., New York (1992) ISBN: 0444-01646-5
12. Agilent: Agilent 4291B RF Impedance/Material Analyzer/Operation Manual, 3rd edn. Agilent Technologies, Tokyo (1999)
13. Palm, W.J.: Introduction to MATLAB 7 for Engineers, 2nd edn. McGraw Hill Professional, New York (2004)
14. Büyüköztürk, O., Yu, T.Y., Ortega, J.A.: A methodology for determining complex permittivity of construction materials based on transmission-only coherent, wide-bandwidth free-space measurements. J. Cement. Concr. Compos. **28**(4), 349–359 (2006)
15. Okubo, H., Shumiya, H., Ito, M., Kato, K.: Optimization techniques on permittivity distribution in permittivity graded solid insulators. In: Conference Record of the 2006 IEEE International Symposium on Electrical Insulation. 11–14 June, pp. 519–522 (2006)
16. Koledintseva, M., Drewniak, J., Zhang, Y., Lenn, J., Thoms, M.: Modeling of ferrite-based materials for shielding enclosures. J. Magn. Magn. Mater. **321**(7), 730–733 (2009)

Printed by Publishers' Graphics LLC
DBT130821.15.15.11